T0134701

Food Microbiology and Food Safety

Practical Approaches

Series Editor:

Michael P. Doyle
Regents Professor of Food Microbiology (Retired)
Center for Food Safety
University of Georgia
Griffin, GA, USA

The Food Microbiology and Food Safety series is published in conjunction with the International Association for Food Protection, a non-profit association for food safety professionals. Dedicated to the life-long educational needs of its Members, IAFP provides an information network through its two scientific journals (Food Protection Trends and Journal of Food Protection), its educational Annual Meeting, international meetings and symposia, and interaction between food safety professionals.

More information about this series at http://www.springer.com/series/7131

Peter J. Taormina • Margaret D. Hardin
Editors

Food Safety and Quality-Based Shelf Life of Perishable Foods

 Springer

Editors
Peter J. Taormina
Etna Consulting Group
Jacksonville, FL, USA

Margaret D. Hardin
Vice President of Technical Services
IEH Laboratories and Consulting Group
Lake Forest Park
WA, USA

Food Microbiology and Food Safety
Practical Approaches
ISSN 2626-7578   ISSN 2626-7586   (electronic)
ISBN 978-3-030-54377-8   ISBN 978-3-030-54375-4   (eBook)
https://doi.org/10.1007/978-3-030-54375-4

This Springer imprint is published by the registered company Springer Nature Switzerland AG
The registered company address is: Gewerbestrasse 11, 6330 Cham, Switzerland

Preface

In the twenty-first century, a growing global population and development in formerly underdeveloped nations has created a demand for food availability in unprecedented amounts and in broad geographical and contextual locations. Such demand means not only that food must be produced with optimal efficiency, but also that technological advances must be utilized to ensure that the food that is produced remains edible and wholesome for the longest amount of time possible in order to enable wide distribution and availability in places where people eat. For instance, the popularity of food trucks, street vendors, and boat vendors has increased globally. The shelf life of food is a key determining factor on how food is distributed, and consequently where and when different food products are available for consumption. Therefore, shelf life must be scientifically determined from both food safety and quality indicators.

Shelf life is determined by several factors, including microbiological, chemical, physical, and organoleptic deterioration. Often these factors are interrelated and interdependent, and the degree of influence of each upon shelf life varies with the type of products. However, for most perishable foods, microbial deterioration is usually the predominant factor. Depending on production process and packaging formats, microbial safety or quality factors can determine the shelf life of a product. Improvements in food processing, distribution, and storage techniques have led to longer shelf life of food products. Examples include the use of antimicrobials and modified atmosphere packaging and to delay or completely inhibit microbial growth for the duration of shelf life. Microbial-based shelf life of perishable foods has long been determined by quantitative plate count data, but now there is much promise in the use of genomics to more fully understand how microbial communities change over the course of shelf life.

This book focuses on microorganisms in perishable food products in keeping with the series on food microbiology and food safety. It does not cover in-depth chemical and physical degradation processes per se, but only as they are a result of microbial growth. The techniques utilized for determination of shelf life, the frequency of shelf life testing for different products, and the interpretation of data to make shelf life determinations receive coverage. We also address newer techniques of analysis of such metagenomics. The reader will gain information on how processing, packaging, and formulation technologies can extend shelf life of perishable foods. Also, insights into the science of shelf life determination of perishable foods is covered including the microorganisms of concern for shelf life determination of major categories of food products.

Jacksonville, FL, USA   Peter J. Taormina

Acknowledgments

The editors thank the following people who have advised or assisted us through the process of planning, writing, and editing this book:
Heather Hart
Jennifer Mason
Michael P. Doyle
Silembarasan Panneerselvam
Susan Safren

Contents

Contributors

Cynthia Ebner Sealed Air Corporation, Charlotte, NC, USA

Margaret D. Hardin Vice President of Technical Services, IEH Laboratories and Consulting Group, Lake Forest Park, WA, USA

Sarah M. Hertrich Food Safety and Intervention Technologies Research Unit, Eastern Regional Research Center, USDA-ARS, Wyndmoor, PA, USA

Ian Jenson Meat & Livestock Australia, Sydney, NSW, Australia

Mandeep Kaur University of Tasmania, Hobar, TAS, Australia

Clyde Manuel GOJO Industries, Akron, OH, USA

Angela Morgan Aptar Group, Crystal Lake, IL, USA

Brendan A. Niemira Food Safety and Intervention Technologies Research Unit, Eastern Regional Research Center, USDA-ARS, Wyndmoor, PA, USA

John Sumner M&S Food Consultants, Deviot, TAS, Australia

Peter J. Taormina Etna Consulting Group, Jacksonville, FL, USA

Steven Tsuyuki Sanitary Design and Corporate Sanitation, Maple Leaf Foods, Mississauga, ON, Canada

Paul Vanderlinde Vanderlinde Consulting, Carbrook, QLD, Australia

Chapter 1
Purposes and Principles of Shelf Life Determination

Peter J. Taormina

1.1 Purposes

Shelf life of food can be defined as the length of time that food can remain in a wholesome, consumable state and retain acceptable quality under normal, expected conditions of distribution and storage. Quality factors can include appearance, odor, flavor, color, texture, nutritional properties, and microbiological populations. These quality factors may diminish at differing or similar rates during shelf life, either remaining within acceptable limit or not meeting quality limits resulting in an unwholesome or nonconsumable state. The extent to which acceptable limits of these factors can be defined empirically for a food product and accurately monitored leads to precision in shelf life designation. Perishable foods require reduced temperature in order to prolong the period for this state of acceptability. Three types of food supply chains have been defined: frozen, chilled, and ambient (Akkerman et al. 2010). The frozen foods' supply chain typically operates at −18 °C, with products like ice cream requiring a frozen chain with an even lower temperature of −25 °C. For the chilled food supply chain, temperatures range from 0 °C for fresh fish to 15 °C for potatoes and bananas, for example. The ambient chain includes products that do not require strict temperature control, such as canned goods. The focus of this book is on the refrigerated shelf life of perishable food products, as that is where microbial growth is of greatest importance.

Perishability is decay, damage, spoilage, evaporation, obsolescence, pilferage, loss of utility, or loss of marginal value of a commodity that results in decreasing usefulness from the original one (Wee, 1993). Certain perishable foods can also support growth of bacterial pathogens during refrigerated storage and therefore have a safety component to shelf life. Shelf life does not necessarily reflect the

P. J. Taormina (✉)
Etna Consulting Group, Jacksonville, FL, USA
e-mail: peter@etnaconsulting.com

© Springer Nature Switzerland AG 2021
P. J. Taormina, M. D. Hardin (eds.), *Food Safety and Quality-Based Shelf Life of Perishable Foods*, Food Microbiology and Food Safety,
https://doi.org/10.1007/978-3-030-54375-4_1

physical state of a product, since many products deteriorate only a while after their shelf life finishes; however, it may reflect its marketable life (Xu and Sarker 2003).

Shelf life determinations have an integral part in perishable food production, processing, sale, distribution, and consumer use. The determination of shelf life of perishable foods integrates food safety and quality factors, including microbiological, chemical, and physical degradation. The sensory properties of foods such as visible appearance, aroma or odor, flavor, and texture are the most obvious aspects of shelf life. Yet the scientific determination of shelf life of perishable foods may also involve tracking the behavior and growth of foodborne bacterial pathogens that do not impart a noticeable sensory change. Ongoing measurement of organoleptic properties of every production lot of food for the purposes of determining or measuring shelf life is seldom feasible for food manufacturers, even if third-party laboratories are employed to perform such work. Therefore, other factors are analyzed, which are indicative of sensory breakdown of products over time such as the increase in microbiological plate counts, increase in oxidative rancidity, or textural changes. To understand the scientific principles of shelf life determination, it is first necessary to outline the purposes for determining shelf life.

1.1.1   Stakeholders

The shelf life of perishable foods is a topic at which business, logistics, science, regulatory compliance, and consumer interests intersect. For each of these stakeholders, the amount of shelf life days assigned to a product is a very important factor that impacts their roles and responsibilities. Commercialization of food products requires collaboration between sales and marketing, business development, operations, and research and development (R&D) functions in order to design food products, scale up prototypes, implement processing and packaging on a larger scale, and produce food products with reasonable consistency. The operations team would be responsible for producing the product in the appropriate manner and under sanitary conditions that result in achievement of the targeted shelf life. Logistics and transportation personnel react to the shelf life of the product and design supply chain and workflows around the number of days allowed. Sales and business development professionals negotiate with customers about the number of days of guaranteed shelf life remaining upon delivery, typically 30–45 days for perishable foods with 3–4 months shelf life. Retail or foodservice end users must utilize the shelf life of the product in the best possible way to maximize sales and minimize waste. The regulator in the production facility or in the marketplace performing inspections takes note of the shelf life of products and responds to consumer complaints. Regulators are also responsible for preventing contaminated or unwholesome food from reaching consumers. Consumers expect a reasonable number of days of useable shelf life and will ultimately validate or invalidate the effectiveness of each of the above by making purchasing decisions en masse.

1.1.1.1 Business Development

Food businesses are keenly interested in food product shelf life for several reasons. Shelf life impacts the logistics and sale of foods because it sets limits on the amount of time for brokering sales of bulk quantities. A short shelf life restricts the production scheduling and impacts labor, raw material supply, transportation, and supply-chain continuity. Fresh cut produce is one such example of a product with a relatively short shelf life. Without antimicrobial interventions, typical fresh cut, bagged vegetables have about a 12- to 15-day shelf life at refrigeration temperatures (Soliva-Fortuny and Martín-Belloso 2003). Many fresh cut vegetable and fruit production plants process, package, and ship on the same day to a distribution center or directly to foodservice or retail customers. If product is shipped to a distribution center, those centers may require 2 or 3 days to receive orders for vegetables and other products prior to loading trucks and shipping to their customers. Allowing for a day or two of secondary distribution, this leaves only about 10 days of shelf life remaining at the point of use or in consumers' possession. This means producers of products with short shelf life have very little leeway and will typically work 7 days a week to fulfill orders. Any disruption in raw material supply, operations, or distribution can severely impact the ability of processors to maintain the supply. Conversely, a long shelf life can allow processors to capitalize on raw material downward price fluctuations and purchase large quantities of raw material when prices are low and store that raw material pending receipt of orders for the product. Long shelf life enables more efficient temporary labor scheduling because food processing plants can schedule long production runs leading to inventory builds. In turn, these long production runs and inventory build-ups enable sales professionals to broker better terms for customers in the marketplace. For example, a producer of cooked ham products may have a validated food safety and quality-based shelf life of 120 days. The processor may find that the raw material, frozen ham muscles, can be purchased on the open market at favorable pricing. At that time, the processor could hire temporary labor to work 14 days straight in order to build a large inventory of fully cooked, vacuum-packaged, ready-to-eat ham with a long shelf life. Further, storage of such products at just above its freezing point can extend the shelf life even further and create more time to find buyers. Under this scenario, the producer can permit necessary downtime in the production schedule, which enables time for preventive maintenance and deep cleaning and sanitizing of the equipment and facility, while the workers can be provided training refreshers and education about health and safety, food safety, and quality. This creates a virtuous cycle, whereby these activities lead to better quality performance, which results in better more consistent product that meets or exceeds the designated shelf life.

1.1.1.2 Logistics

Well-managed food logistics can reduce the incidents of product reaching the end of shelf life without a buyer and without reaching a consumer. The perishability of fresh food products limits the opportunities to use inventories as a buffer against variability in demand and transportation (Ahumada and Rene Villalobos 2009). Food processors must avoid over-ordering raw material supplies and packaging materials and overproducing finished product beyond the actual demand. This presents a challenge to forecast accurately the demand. Conversely, retail or foodservice customers must avoid overestimating consumer demand and thereby purchasing too much product. Thus, efforts to successfully extend shelf life of perishable foods can create more flexibility in the supply and demand planning, operations, and transportation dynamic, as well as potentially provide customers more time to display and potentially sell food products.

Designing food distribution systems that are able to provide high-quality food in a cost-efficient way is a challenge that requires collaboration between cross-functional teams such as logistics, food engineering, and operations management (Apaiah et al. 2005). Food produced by manufacturers, packers, and processors may be shipped directly to retail or foodservice points or is often sent to distribution centers that serve as intermediaries in logistical transport and delivery of food (Fig. 1.1). While this general distribution structure remains the norm for a majority of food, increasingly, food is being sent directly to consumers as online meal kit delivery services rise in popularity. Direct-to-consumer food distribution models drastically change the shelf life potential due to lack of consistently verifiable temperature control.

The limited shelf life of perishable food products, requirements for strict temperature and humidity control, possible interaction effects between products, time windows for delivering the products, high customer expectations, and low profit margins make food distribution management very challenging (Akkerman et al. 2010). On a strategic level, perishability may also play an important role by forcing better relations and more integration between the supply chain networks of within and between organizations (Amorim et al. 2013). For example, most of the food industries rely on third-party logistic providers (3PL) to perform the distribution of their products. A good relation between those companies can result in additional gains in efficiency leading to lower costs of distribution and less waste. At the other end of the supply chain is a vendor-managed inventory system, relating a supplier of a perishable raw product and a company that processes this product.

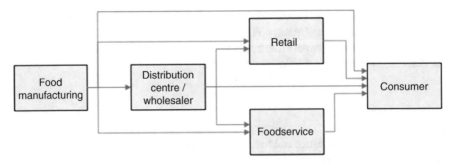

Fig. 1.1 General structure of distribution within the food supply chain (from Akkerman et al. 2010)

Vendor-managed supply chains mean perishable foods produced on demand factoring in expected shelf life. In either system, enhanced communication between these entities that enables better supply and demand planning should potentially lead to less spoilage and, hence, less costs that all parties may reap, including the consumer.

A forecast-based ordering model for Nordic supply chains for milk, fresh fish, and poultry were studied by Kaipia et al. (2013) as a means to reduce waste and increase sustainability. They concluded that supply chain structure should be streamlined to avoid additional handling and delays and that the order penetration point (OPP) location needs to support the specific features of fresh food supply chains. Their initial findings indicated that the OPP should be moved as close to the retail customer as possible. They also determined that an efficient forecasting process becomes more important in this model because a larger share of the chain operates on the basis of forecasts. The demand data should be utilized to balance operations between made to order (MTO) and made to stock (MTS) production. They concluded that demand data form a good basis for forecasting and, on the other hand, can trigger larger MTO production instead of producing to stock. The way to utilize demand data needs to be selected in a way to support optimally the specific product characteristics and demand patterns (Kaipia et al. 2013). Perishable products characterized by a short shelf life are well suited to the MTO model, especially if raw materials can be on hand and ready for processing into finished product. For foods like fresh fish or fresh cut produce, timing a harvest date to on-demand production could be a more significant challenge.

Utilization of information technology can potentially enhance the efficiency of food logistics. Radio frequency identification (RFID) has been researched as a means to manage highly perishable food inventory distribution (Grunow and Piramuthu 2013). When a food item is RFID-tagged, the item's remaining shelf life is known with certainty to the retailer, distributor, and customer, and the retailer cannot sell an item beyond its "expiry date." While RFID temperature and time monitoring in trucks are not uncommon, monitoring at the pallet level is costlier, and monitoring of individual units is cost-prohibitive in most cases. However, even pallet-level RFID tags with sensors could enable distributors to send those pallets with higher remaining shelf life to the more distant retailers. The value of real-time access to product shelf life information will be realized by quality and food safety professionals responsible for assuring safe, wholesome food is delivered and consumed within the code date.

Technologies like Blockchain are now being used to facilitate the transactional transfer of food through the supply chain (Tian 2016, 2017). The efficiencies gained by use of Blockchain can benefit perishables with a relatively short shelf life. Blockchain is now being used to establish within the supply chain a permission access-based, viewable, and immutable history of network, which can be shared among all nodes in the system (Tian 2017). With such incredible insight into the history of food, there will be more insight into the actual shelf life of food products for brokers, customers, and consumers.

1.1.1.3 Science

Shelf life determination must be science-based and supportable using data. The limitations of number of days of shelf life are a matter of what can be established with scientifically accepted methods (Kilcast 2001; Man 2004). The development of new technologies to prolong the shelf life of foods include improvements to raw material harvesting and speed to processing, advanced food processing techniques, packaging technology, formulation, and cold-chain integrity. Indeed, shelf life extension has become a goal of many food producers in order to alleviate some of the previously discussed business challenges with producing and distributing perishables. Consumer and customer demands for "clean label" food and beverage products have resulted in the expectation that traditional preservatives like benzoate, sorbate, nitrite, and propionate (Brul and Coote 1999) be removed and replaced with ingredients that are perceived as natural such as cultured antimicrobials, vinegar, bacteriocins, and plant extracts (David et al. 2013). The resulting span of shelf life of so-called clean label products is seldom the extent possible by incorporation of traditional preservatives. Scientists and research and development professional must study these food systems and challenge them with worst-case conditions of processing, packaging, storage, and use and must prove food products will hold-up for the designated shelf life.

The research, development, and technology team of a food company or external partners in academia or food laboratories could employ culinary and food science techniques to create a product that meets the market demand but must also understand the sales and marketing professional has a real need for competing in the marketplace. This means that the product developers, processing engineers, packaging engineers, and quality and food safety professionals must work together to develop least cost options for formulation, process, package, and shelf life. The shelf life has to be sufficient to enable production planning professionals to meet the supply demand and for logistics professionals to deliver product to the customer in a timely manner. The sales, marketing, and business development professionals work with research and development scientists and technicians to craft a product that meets the market needs. The formulation and processing techniques required to meet shelf life targets must support the business and marketing objectives in terms of costs of production and in nutritional labeling and ingredient usage that meet consumer expectations. Operations personnel manufacture product according to the specifications set by R&D professionals. R&D professionals design formulations and usually outline target processing parameters for achievement of a desired shelf life. Quality personnel are responsible for monitoring product quality and performance and may utilize ongoing shelf life testing of representative samples of production as one of a variety of metrics, albeit a lagging one. Growth of spoilage microorganisms during shelf life can result in the development of an unwholesome product over time, and therefore food safety and quality professionals and regulators are concerned with the scientific aspects of shelf life determination. Pathogen growth during refrigerated shelf life can cause foods to become unsafe for consumption, and regulations exist concerning the suppression of growth of psychrotrophic or psychrotolerant bacterial pathogens like *Listeria monocytogenes*, non-proteolytic *Clostridium botulinum*, and *Bacillus cereus* (U.S. Department of Agriculture 2015; U.S. Food and Drug Administration 2017). R&D professionals oversee the performance of challenge studies to validate safety of perishable food products throughout shelf life.

R&D professionals are often tasked with the job of collecting shelf life validation data and presenting it to foodservice or retail customers for approval. Foodservice and retail customers review and accept shelf life validation data provided by their various manufacturing suppliers, and if such data are deemed acceptable will allow these perishable food products into their inventory. Retailers must have sufficient time to distribute the product through their network and to display the product to the consumer. The consumer also must have a reasonable and acceptable amount of time to see the product package in the retail environment, and store, use, or reuse the product once it has been purchased.

Martins et al. (2008) reviewed computational methods for shelf life dating. Computational methods use modeling techniques that bridge experimental data such as quality and safety characterizations, stress level testing, accelerated shelf life testing, and distribution chain data with information systems. This enables multi-scale scenario analysis and interpretation in order to better understand the complex food quality dynamics of shelf life of food. Certainly, a scientific process of experimental data collection about the product coupled with real distribution information offers promise as a means to accurately predict shelf life and set accurate code dates.

1.1.1.4   Regulatory

Control of pathogen growth is often the most necessary part of shelf life determinations and can be the limiting factor in setting shelf life of certain perishable foods. Several foodborne pathogens have little to no bearing upon shelf life of refrigerated foods due to the fact that they are mesophilic and growth is inhibited at refrigeration temperatures. For example, pathogens within the family *Enterobacteriaceae* such as *Salmonella* and pathogenic *Escherichia coli* are typically not able to grow at refrigeration temperatures (Roberts and Tompkin 1996). Even shelf stable products that might support growth of xerotolerant yeasts and molds would not support growth of most bacterial pathogens. Growth of psychrotrophic pathogens such as *Listera monocytogenes*, *Yersinia enterocolitica*, and group II *Clostridium botulinum* must be assessed on refrigerated perishable foods. These psychrotrophic bacterial pathogens of concern are reviewed in detail in Chap. 3.

Microbial spoilage of food and beverage products can be caused by a number of factors, such as a loss of process control, post-processing contamination, inadequate packaging performance, damage during distribution, or temperature abuse during distribution, storage, or display. The causative microorganisms of food and beverage spoilage are usually not the same as those that are attributable to foodborne illness. However, in some instances in North America, spoiled products have been subjected to class II recalls (United States Department of Agriculture, Food Safety and Inspection Service 2015), rather than a more discreet market withdrawal. With the recent mandatory recall authorization provided to the U.S. Food and Drug Administration in the Food Safety Modernization Act (Anonymous 2011), the difference between "market withdrawal" and "recall" is an important regulatory and legal matter that impacts food brands and the bottom line for food businesses. Psychrotrophic lactic acid bacterial growth and spoilage effects were responsible for numerous recalls in Belgium of meat, dairy, vegetable, egg products, and composite foods (Pothakos et al. 2014). Thus, spoilage and shelf life do become a matter of regulatory concern.

The term "undesirable microorganisms" as defined in the Hazard Analysis and Risk-based Preventive Control Final Rule includes those microorganisms that are of public health significance, that subject food to decomposition, that indicate that food is contaminated with filth, or that otherwise may cause food to be adulterated (U.S. Food and Drug Administration 2015). Spoilage of food prematurely before the expiration date results in undesirable if not unwholesome food. However, questions remain as to what circumstances and which microorganism-product interactions turn a food spoilage event into a food safety concern. Consumers are also confused about the differences between true food safety hazards and undesirable, but innocuous, food spoilage. Expert interpretation of the microbiological data, risk assessments on the product, product status in the marketplace, and normal consumer use and handling of the product must be evaluated on a case-by-case basis to make decisions about potential public health impact of spoiled foods implications on market withdrawal or recall.

1.1.1.5   Consumers

Consumers form quality expectations of food products based on perceived cues prior to purchase, and post-purchase quality experiences will impact future purchases (Grunert 2002). At retail markets, consumers primarily observe the visual appearance of products and assess potential quality and solely this way for many packaged (sealed) products. Consumers expect a reasonable number of days to store and use perishable food products once purchased (Man and Jones 1994). Consumers certainly use the sense of smell to assess food quality whenever that is possible. Package labeling also weighs heavily into consumer perception of the food. Consumers interpret the printed code date, or expiration date, on the package with some flexibility, with variation depending upon the type of food product (Van Boxstael et al. 2014). Consumers are less likely to eat expired raw products than expired ready-to-eat products.

In terms of different *date labelling*, likelihood of consumers checking expiration dates appears to depend on the food category (Aschemann-Witzel et al. 2015). Consumers check expiration dates frequently for products in which a decrease in quality is risked and for products with which they have experience of usage (as measured by the household consumption rate for the category). Relatedly, the willingness to pay for a perishable product decreases throughout its shelf life. Freshness dating influences the acceptability of products in a discontinuous or non-linear manner because it influences perceptions of freshness and of healthfulness, not of safety (Wansink and Wright 2006). As a product approaches its "best if used by" date, there may be more for a manufacturer to lose than to gain by having decided to use "freshness dating" in the first place.

1.1.2 Assuring Quality and Wholesomeness of Food through Code Dates

A code date is the date stamped on the product box and/or package that indicates to the end user or intermediate user, the time for which the product can be safely and acceptably used. Code dates vary with the product type and the intended customer. For example, food products sold to foodservice distributors will typically end up in institutional or restaurant kitchens. In those settings, food preparers and chefs will likely see boxes and packages stamped with date codes that are somewhat cryptic. This is referred to as *open dating*. Some date codes will indicate only the date of production. Other date codes will be Julian dates. Julian dates (abbreviated JD) are simply a continuous count of days and fractions since noon Universal Time on January 1, 4713 BC (on the Julian calendar). Sometimes Julian dates are nested within other codes that could relate to production line number, facility, or customer. Foodservice workers therefore have considerable discretion about how long to use such products, whether they are held frozen or refrigerated. If frozen product is received, foodservice workers must decide how long the product is usable once thawed, unless guidance has been provided by the manufacturer. In some cases, government entities provide safe handling days for products. At the retail point of purchase, consumers are presented with a variety of formats through which the shelf life is communicated (Newsome et al. 2014; Wansink and Wright 2006). This has become a point of confusion and misunderstanding with regard to safety and whole-someness of foods in general, but it is of more importance with perishable foods compared to shelf-stable foods due to the potential for microbial growth during shelf life.

1.1.2.1 Use by, Sell by, Best by, Best Before, Best If Used by, and Enjoy by

Code dates formats at the retail point of purchase have well been researched. In a Belgian study, it was found that 30.4% of the respondents declared they did not know the difference of the meaning between the best before and use by date and only about half of the respondents indicated to use the type of label (difference between the labels) for assessing the edibility of a product (Van Boxstael et al. 2014). The authors surmised that part of the confusion lies in the fact that products that have only quality degradation as the limiting factor (e.g., mayonnaise) and perishable products that have safety and quality factors determining the shelf life (e.g., meats), both have *use by* dates. They proposed to simplify this by having government assign all refrigerated products a *use by* label (and the communication that *use by* products are unsafe to consume after the shelf life date and that for *best before* products the edibility can be judged by the consumer). The authors admitted the difficulty in predicting the outcome of this simplified approach and noted there to be advantages with respect to food safety and a mixed outcome with respect to food waste.

The issue of date labelling was found to be a primary issue in consumer-related food waste (Aschemann-Witzel et al. 2015). The terminology and use of date labels vary widely, and this contributes to consumers' confusion about the meaning (Newsome et al. 2014). A more uniform and consistent date label display format that gets away from *sell by* or *display by* has been proposed to reduce misunderstanding for consumers (Aschemann-Witzel et al. 2015; Newsome et al. 2014).

For perishable foods, consumers search for visual and other cues of freshness, such as expiration dates. The greater the risks associated with a product, the more frequently consumers check expiration dates (Tsiros and Heilman 2005). If consumers perceive risk associated with the quality of a perishable food product as it approaches its expiration date and foresee possible health risks from consuming the product, they are more likely to check and adhere to expiration dates. Consumer willingness to pay (WTP) for perishable foods decreases markedly as products approach expiration dates (Fig. 1.2). WTP for chicken and lettuce declines rapidly even 6 days preceding expiration (Tsiros and Heilman 2005).

Expiration of perishable food accounts for about 20% of the total unsalable items in grocery stores, drug stores, and wholesalers in the United States (Joint Industry Unsaleables Leadership Team 2008). From 2005 to 2007, 56% of retailers and distributors surveyed indicated that unsaleable costs due to expired products have increased. The report indicated that since the use of open date coding is diminishing and more intuitive code date formats such as *use by* are more prevalent, consumers are more aware of expiration dates, and therefore, manufacturers are more conservative about shelf life days assigned to products. This has led to more product expiring in the marketplace.

1.1.2.2   Food Waste

Expired products (past the *sell by* date) account for much of the losses at retail supermarkets. It has been estimated that 12% of fruits and vegetables and 9.5% of seafood is discarded at North American retail markets, and consumers discard nearly 30% of seafood and produce (Gunders 2012). The most common reason for waste at the retail store is that products' expiry dates have passed (Kaipia et al. 2013). More than a quarter of food discarded from retail stores in Austria were had no defect other than the *sell by* date had passed (Lebersorger and Schneider 2014). The causes for this include ordering more than real demand or products reaching the

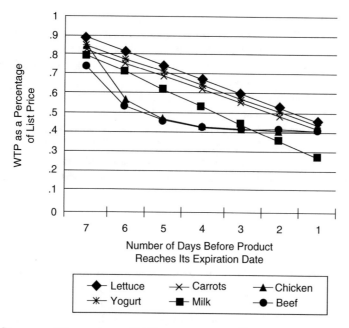

Fig. 1.2 Consumer willingness to pay (WTP) as a percentage of list price throughout a food product's perishable shelf life (from Tsiros and Heilman 2005 with permission)

store shelf too late and with a short remaining shelf life (Mena et al. 2011). Therefore, inefficiencies or disorganization in perishable food supply chains can directly impact the amount of food that is expired before consumption and ends up being destroyed.

1.1.3 Expiration = Decision Time

1.1.3.1 Distressed Product

When food in distribution centers or cold-storage warehouse facilities near the end of the stated shelf life, it is too late to change the code date. Often, product managers or sales and marketing professionals might ask the scientists for extension of shelf life after the product has been produced, packaged, shipped, and received. Apart from the appearance of impropriety, there are inherent risks with addition of days to product shelf life ex post facto. The proper food safety and quality culture should influence the decision by the manufacturer to hold fast to the shelf life of the product and not allow additional days to be added to the code date. This pertains to refrigerated, perishable foods, but the concept is important for shelf stable foods and frozen foods as well. Even if those products do not deteriorate at a rate that would greatly impact the addition of a few weeks of shelf life, the very practice of granting extensions can lead to dangerously false perceptions that quality and food safety is negotiable, and allowance of other unacceptable quality practices. It could eventually overwhelm the technical quality assurance personnel, as they will be constantly asked for approvals of extensions of shelf life. Rather, the safest, most appropriate approach is to perform proper shelf life testing (i.e., validation) to establish the maximum reasonable shelf life to assign to products. Inventory management systems can alert product managers, sales, and marketing professionals when food product lots are close to reaching the expiration date. At such point, product can be allocated as "distressed," signaling to the business that it is time to unload this inventory customers at a discount or risk not selling it at all. Strong inventory management systems should include an ongoing plan for discounted selling of distressed product.

1.1.3.2 Donation

Donation is the second option when food is near the expiration date. If a manufacturer or distributor or retailer possesses food that is close to the expiration date, and the product cannot be sold, donation is a sustainable option. In the U.S., the Bill Emerson Act of 1996 (Anonymous 1996) encourages the donation of food and grocery products to nonprofit organizations for distribution to needy individuals by giving the Model Good Samaritan Food Donation Act the full force and effect of law. The law states that "a person or gleaner shall not be subject to civil or criminal liability arising from the nature, age, packaging, or condition of apparently wholesome food or an apparently fit grocery product that the person or gleaner donates in good faith to a nonprofit organization for ultimate distribution to needy individuals." There are a variety of organizations that receive and redistribute food donations on a large scale, and so the option for donation of nearly expired food should be a forethought for food producers, distributors, and retailers.

1.1.3.3 Diversion

Diversion is another possible choice for food close to the expiration date. Food products can be diverted as an incoming raw material stream to other forms of food processing, if it is compatible. This is an option for raw commodities, such as raw meat being diverted to a cooking process, more so than for packaged foods. Besides compatibility, feasibility becomes a factor as product is often in boxes with packaging material. Businesses must consider the labor and other costs associated with reprocessing and/or repackaging nearly expired food to divert it to another product stream. Another option is to divert to animal food production. However, animal food is somewhat precise in terms of desired proteins, carbohydrate, fat, and nutrient content. Further, the food ingredients used for human food are not always approved for animal food.

1.1.3.4 Destruction

Destruction should be the last option in dealing with expired food. It is an unfortunate and sometimes unavoidable consequence of the food business. Sustainable companies should predefine the proper channels for food donation and diversion, and once those channels are exhausted, follow a destruction plan that does not ignore sustainability. Companies should create operating procedures to assure removal of packaging materials such as pallet wrapping, corrugated boxes, plastic food packaging so that it may be recycled and so that the perishable food matter is separate and the only part that is either incinerated or sent to a landfill. Ideally, unless the food producer is at fault, it should not be always responsible for the entire cost of destruction. In instances when product must be destroyed due to the end of shelf life, a shared cost would be a reasonable way to assure that it is a last resort, and that neither producer, broker, nor customer are compelled to force destruction before a reasonable consideration of the other potential options.

1.1.3.5 Scandal

Unfortunately, incidents of food fraud concerning shelf life have occurred. The owner and a top executive of US egg company were found guilty of distributing adulterated eggs and were sentenced by the federal government to prison time and fines (U.S. Department of Justice 2015). Among the various violations, they were found guilty of instructing employees affix labels to egg shipments that indicated false expiration dates with the intent to mislead state regulators and retail egg customers regarding the true age of the eggs.

In Huanan province in China, several gangs smuggled frozen meat that was so old that it was dubbed, "zombie meat" since it was up to 40 years old (Alice 2017). The rotten meat included pork's feet and chicken's claws, chicken's wings and other meat products were moved to the mainland via Vietnam, with smugglers hiring residents of border areas to move the products to Chinese border cities and then on to Changsha before the products were transported to several sites within China. Importers soaked them in hydrogen peroxide, a banned food addictive, to make them look healthy and fresh and to extend their shelf life (Alice 2017). Global, publicly traded quick-serve restaurant chains were involved in this scandal, which led to substantial loss in share price. This and other non-shelf life-related food safety scandals in China prompted a comprehensive revision of the 2009 Food Safety Law of China on April 24, 2015 (Geng et al. 2015).

Scandals related to shelf life fraud such as these are likely to diminish because modern food and beverage production and brokering of food commodities have reached unprecedented levels of transparency. Consumers have increasing concerns about the safety, quality, and authenticity of foods as a result of various food scandals, recalls, and outbreaks (Bánáti 2011; de Jonge et al. 2010). These concerns have translated to the need for more visibility into food and agriculture production systems, between businesses, and between business and consumers. Among those concerns, the shelf life of foods has emerged as one of the key areas. With the intense scrutiny of food safety and quality systems through auditing and quality data management and sharing systems, gone are the days when shelf life of perishable foods can be arbitrarily set. Further, the unscrupulous practice of deeming shelf life longer than technologically supported will no longer go unnoticed and without consequence. Therefore, determination of shelf life of perishable foods must be a scientific process following conventional principles.

1.2 Principles of Shelf Life Determination

1.2.1 Quality Deterioration Rates

The shelf life of foods is *simply* the duration of time from the point of production that the product is safe, wholesome, and suitable for consumption. Determining those factors, however, is not simple due to the variety of intrinsic properties that interactively affect shelf life performance and the high likelihood of variability in the external environment of temperature, humidity, light, and time (Office of Technology Assessment 1979; Labuza 1984; Fu and Labuza 1993). Shelf life of perishable foods is generally based upon the rate of microbial, chemical, and physical degradation, which work individually or, more often, in concert to speed the rate of organoleptic deterioration of the product. Of each of the external factors, temperature has the greatest impact on the rate of deterioration of most perishable foods. In order to understand the quality deterioration of foods, a few mathematical equations and concepts are presented, but this chapter does not provide a full picture of the quality deterioration reaction kinetics and equations. A full review of such information can be found from various other sources (Office of Technology Assessment 1979; Singh 1994; Singh and Cadwallader 2004; Van Boekel 2008; Labuza 1984; Fu and Labuza 1993). A brief overview of quality deterioration reaction kinetics gleaned from these sources will be presented here in order to introduce the concept of quality and shelf life.

The decrease in a quality attribute of food during storage can be defined in a first-order reaction. The temperature-dependent general loss of food quality can be represented by a mathematical Eq. (1.1):

$$\text{Rate} = \frac{dA}{d\theta} = kA^n \tag{1.1}$$

where:

A = the quality factor to be measured

θ = time

k = a constant that depends on temperature and other factors (e.g., water activity, pH)

n = a power factor called the order of the reaction, which defines whether the reaction rate is independent of the amount of quality remaining

$\dfrac{dA}{d\theta}$ = the rate of change of A with time. A negative sign is used if the deterioration is a loss of A, and a positive sign is used if it is for production of an undesirable end product of deterioration.

This general loss of quality equation results in a rate calculation, but in practice, the results of shelf life studies are rather defined as the amount of A left or produced as a function of time.

Based on Eq. (1.1), the order of the reaction (n) is often treated as if it is equal to zero. This zero-order reaction assumes that the rate of loss is constant for some constant temperature:

$$\text{Rate of loss} = \frac{dA}{d\theta} = k = \text{Constant at some temperature} \tag{1.2}$$

If that equation is integrated, it becomes Eq. (1.3):

$$\text{Amount left}\left(A\right) = \text{Initial amount}\left(A_0\right) - k\theta \tag{1.3}$$

where:
  A_0 is the initial (time zero) value of the quality factor.
  To put this in terms of shelf life, the equation becomes (Eq. 1.4)

$$\theta_s = \frac{A_0 - A_s}{k} \tag{1.4}$$

where:
  θ_s is end of shelf life in time (in days, weeks, and months)
  A_s is the maximum allowable loss value, or the threshold point at which quality has deteriorated such that the product no longer has acceptable quality.
  The rate of deterioration becomes the rate constant in Eq. (1.5):

$$\text{Rate of quality loss} = k = \frac{100\%}{\theta_s} = \text{Constant\% loss per time} \tag{1.5}$$

This zero-order reaction means that n from Eq. (1.1) must equal 0. The constant, k, could be different factors such as constant rate of development of oxidative rancidity or vitamin loss. However, not every quality factor will deteriorate at a constant rate, such as described by a zero-order reaction. Particularly, microbial growth tends to follow a first-order, or exponential, reaction for which $n = 1$. Mathematically, the rate of loss for first-order reactions is shown in Eq. (1.6):

$$\text{Rate of loss} = \frac{-dA}{d\theta} = k(A) \tag{1.6}$$

Integrating Eq. (1.6) gives a logarithmic function of the first-order as shown in Eq. (1.7):

$$1_n = \frac{A}{A_0} - k\theta \tag{1.7}$$

Graphically, the zero- and first-order reactions with respect to shelf life of foods are shown in Fig. 1.3.

Shelf life limit (θ_s) would be set at the point when the zero-order reaction (linear decrease in quality over time) crosses the A_s. However, as mentioned many deterioration scenarios follow the first-order reaction rate, and sometimes the A_s will not be reached graphically as the quality tails off asymptotically. Rather than claiming a length of shelf life days far beyond where that curve may extrapolate, a more reasonable choice is to plot the quality factor as semi-log and select a higher A_s (i.e., a higher quality standard) such that the first order quality deterioration curve plotted logarithmically crosses A_s at some θ_s as shown in Fig. 1.4.

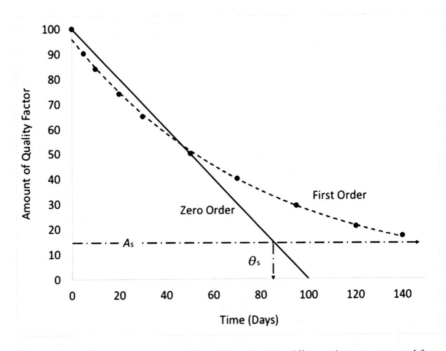

Fig. 1.3 Loss of food quality as a function of time showing difference between zero- and first-order reactions. End of shelf life time (θ_s) occurs when quality reaches maximum allowable loss value (A_s). (Adapted from Labuza 1984, with permission)

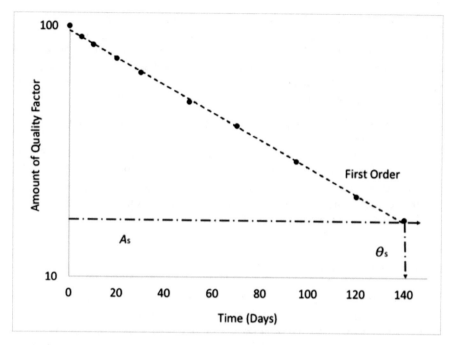

Fig. 1.4 Loss of food quality for a first-order reaction plotted as semilog. In this plot, the end of shelf life (θ_s) occurs when quality reaches maximum allowable loss value (A_s). (Adapted from Labuza 1984, with permission)

The term A_s could also be considered a percent of usable quality remaining. The difficulty is in designating the criteria for A_s. This can be a complicated quality metric based upon sensory research, or it can be an arbitrary value based upon populations of aerobic mesophilic microorganisms measured as the aerobic plate count (APC) and/or lactic acid bacterial counts ("Lactics") as measured by anaerobic, mesophilic incubation on specific media. If only one factor can be used, it is best to select the most pertinent quality metric that degrades the fastest. It may very well be that for some products, APC or Lactics are indicative of organoleptic degradation and would be ideal for A_s.

1.2.2 Defining Quality Factors

Food are inherently unstable, as each type of quality deterioration reaction will proceed at some rate affected by intrinsic properties of the food, as well as the extrinsic factors like light, temperature, humidity, and atmosphere. Food quality is what is usually measured to determine the shelf life of foods, but the safety factors like bacterial pathogen growth over time at the given storage temperature, take precedence if they are relevant. In food product development, the quality metric must first be defined for the food product based on specifications of microbial, chemical, physical, and organoleptic attributes. The measurement of the quality of the food pertains to the degree to which it meets these predetermined attributes. The ultimate shelf of food is influenced by what the food composition is, how the food is processed, how it is packaged, and how it is stored. Indicators of food quality change over time and can be classified in four ways (Van Boekel 2008):

- Chemical reactions, principally from oxidation or Maillard reactions
- Microbial growth, leading to spoilage and, if pathogens grow, unsafe food
- Biochemical reactions: endogenous enzymes that catalyze reactions leading to quality loss (enzymatic browning, lipolysis, proteolysis, etc.)
- Physical reactions: such as moisture migration leading to staling, softening, or "freezer burn"; loss of heterogeneity of dispersed or suspended particles leading to coalescence, aggregation, and sedimentation

1.2.2.1 Microbiological Profiles

Decades prior to the time of this publication, little was known about the relationship between microbial activity and biochemical spoilage parameters of food under different packaging and storage conditions (in't Veld 1996). Back then, the ability to characterize the total microflora and metabolites developing during food spoilage, but the ability to identify specific microorganisms in relation to food composition was lacking (in't Veld 1996). Now metagenomics techniques coupled with bioinformatics have been applied to food shelf life research (Benson et al. 2014). Quantitative PCR (qPCR) of 16S rRNA can be used to confirm dominant and subdominant species of bacteria that are contributing the most to the overall microbial population, as well as understand the role of bioprotective cultures in food systems and investigate how environmental processing conditions impact these microbial communities (Cauchie et al. 2017; Rouger et al. 2018). Advanced molecular techniques are beginning to be applied within the food industry, but the standard plate count techniques are by far more utilized and will continue to be into the future; plate count limits are still ingrained in specifications within the food business and in regulatory inspections.

Setting the minimum acceptable quality of foods requires attention to the scientific requirements of maintaining wholesomeness, as well as regulatory limits and customer required limits. For example, there may be a customer limit for Aerobic Plate Count of 20,000 CFU/mL, and so that it would be at least one criterium for end of shelf life. Microbial degradation at refrigeration temperatures is commonly the limiting factor for shelf life as byproducts of microbial metabolism alter the physiology and integrity of food systems (Gram et al. 2002; Remenant et al. 2015). As mentioned previously, microbial growth curves follow first-order reaction kinetics. Predictive microbiology models can account for microbial ecology and physiology to accurately estimate the remaining shelf life of such food systems (McMeekin and Ross 1996). One such example shown in Fig. 1.5 is the growth of *Photobacterium*

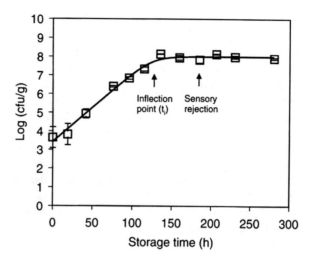

Fig. 1.5 Growth of *Photobacterium phosphoreum* in naturally contaminated cod fillets stored at 0 °C in an atmosphere with 100% N_2. Error bars indicated standard deviations ($n = 2$). From Dalgaard et al. (1997) with permission

phosphoreum in cod fillets stored at 0 °C (Dalgaard et al. 1997). Growth of the spoilage bacterium was predicted and associated with organoleptic failure, which goes beyond the arbitrary plate count limit of quality and shows how populations of a specific spoilage bacterium correspond to real degradation. Shelf life prediction based upon microbial growth at non-isothermal conditions simulating distribution and storage variation has been researched for a wide variety of perishable foods including fish (Koutsoumanis 2001) and ham (Kreyenschmidt et al. 2010) for example. There have been many more applied research studies concerning predictive microbial modeling for shelf life prediction.

1.2.2.2 Consumer Acceptance

There are different levels of quality, and interpretation of product quality involves laboratory techniques, as well as sensory science. Otherwise, quality becomes amorphous and/or subjective. While consumers are considered the most appropriate gauge of determining the shelf life of food, assemblage of consumer panels for multiple measurements necessary for shelf life studies is not feasible (Gámbaro et al. 2004). Nonetheless, the performance of consumer sensory analysis can be achieved for larger product groups, and data resulting from such work can form the basis for shelf life prediction for product categories. For example, during the product development phase for a line of Greek-style yogurts, scientists should conduct consumer acceptance testing on a model product formulation to establish the likely shelf life of the product category. With such data in hand, future product variations (e.g., flavors, packaging types, portions) could reasonably be expected to perform similarly over time. Consumer preference of one product over another is an important determination that impacts the shelf life of products. Depending on the product attributes, as the food ages, the degree of consumer acceptance can diminish. So, a decision must be made as to the acceptable quality attributes of the product during shelf life. Consumer acceptance of food products varies with individuals, and so the shelf life of foods depends not only on the food quality but also on the interaction of the food with the consumer (Hough et al. 2003).

Survival analysis is a branch of statistics that has been applied to the study of the shelf life of foods (Hough et al. 2003; Gámbaro et al. 2004). The focus of survival analysis is on the risk of consumer rejection of the product and that impacts on shelf life, rather than on the product deterioration. Another quality designation that helps determine the shelf life of products is the degree to which a product meets customer and consumer expectations at a base level. Determination that a product is suitable or fit for purpose means that the product meets the minimum standards. The "Just Noticeable Difference" (JND) concept applies to sensory evaluations performed by trained panel and/or consumers. The shelf life may also be limited by consumer complaints of some factor pertaining to not only the product itself, but also the way the product performs in the package on display.

1.2.3 Shelf Life Testing

Assigning a shelf life to a product involves the research and development process of formulation, processing parameters, packaging, and consumer acceptance testing. The deterioration of food can be due to microbial, chemical, or physical factors, and all of these can singly or in combinations affect the organoleptic properties of the product. Organoleptic attributes of appearance, odor, taste, and texture are each important and measurable. The factors that impact shelf life can be safety related, such as growth of foodborne pathogens over time, or can be quality related, such as growth of spoilage microorganism. While food safety implications are the limiting factor in many perishable foods, the ways this is determined are much different than the quality factors.

Shelf life testing occurs at various stages and involves differing approaches at each stage of a product life cycle (Table 1.1). The process of establishing a product shelf life occurs in the new product development phase or the research and development phase. New product development entails something altogether new and different from existing product lines. Such novelty necessitates relatively elaborate shelf life testing in order to determine the usable days. Challenge studies are an important aspect of measuring the potential for pathogen growth during shelf life (National Advisory Committee on Microbiological Criteria for Foods 2010) and should be conducted early enough in the process in order to obtain results prior to decisions to launch new products. Routine shelf life testing involves collecting freshly manufactured food products in their finished, packaged state, and storing them at a defined refrigerated temperature and periodically analyzing samples for some of the aforementioned factors. Changes such as microbial growth, oxidative rancidity development, organoleptic degradation (e.g., color, flavor, texture), or increase in purge are monitored over time until thresholds of acceptance are met or exceeded. At that

Table 1.1 Types of shelf life testing different stages of food product development and impact of data generated

Stage of product development	Type of shelf life testing	Use of data
1. Ideation/proof of concept	Predictive microbial modeling	Go/no go decisions
2. Prototype development	Chemistry (proximates, pH, a_w); predictive microbial modeling	Attain food safety and quality approval to develop further
3. Pilot plant testing	Shelf life testing at refrigeration and moderately abusive temperature; measuring general microflora, fat oxidation, texture, and organoleptic acceptance; microbial challenge study[a]	Attain food safety and quality approval for plant production; set shelf life target days (i.e., code date)
4. Scaled-up plant trials	Shelf life testing at refrigeration temperature; measuring general microflora, fat oxidation, texture, and organoleptic acceptance; microbial challenge study[a]	Food safety and quality validation
5. First production run	Shelf life testing of representative samples; samples stored at refrigeration and analyzed for organoleptic, general microflora, fat oxidation	Quality validation
6. Routine production	Shelf life testing of representative samples retrieved at a necessary frequency in order to assess quality; samples stored at refrigeration and analyzed for organoleptic, general microflora, fat oxidation	Quality monitoring

[a]Microbial challenge testing will take 1.25–1.5 times the shelf life target. Therefore, challenge study results for a product with 30-day target shelf life would not be completed until up to 45 days. For products with a long, refrigerated shelf life (e.g., 120 days), there may be a need to perform a challenge study earlier in the process, such as at the pilot plant stage

time, a shelf life test is deemed complete, and the number of days the product was acceptable is considered the result. Alternatively, a pass or fail rating is given, and those results are aggregated over time. Routine shelf life testing involves no introduction of other microorganisms to the food product. It is simply measuring the behavior of the autochthonous microflora over time, along with the other factors.

Shelf life testing is beneficial in the sense that it provides an actual profile of the product as it changes over refrigerated storage. It is limited in the sense that the level and type of autochthonous microflora present from production lot to production lot can vary. Also, there are sometimes variations between samples collected from one particular lot. Due to cost, limited refrigerated storage space for samples, and laboratory labor required, shelf life testing is usually performed only periodically and often involves analysis of only one packaged unit per time point. Indeed, if duplicate or triplicate samples are not analyzed at each time point, the sample variability can lead to inaccurate results and conclusions. It is often the case that one packaged food product sample out of several will succumb to spoilage days or weeks before the other samples from the same lot exhibit signs of degradation. Data in Table 1.2 would indicate that everything is performing adequately. The starting populations are low as expected, and even though the T14 sample shows markedly higher counts for both APC and Lactics, the subsequent sample at T28 is comfortably back into the acceptability limits. The final analytical unit sampled at T60 indicates that populations were approaching the limit but were still sufficiently below thresholds. Therefore, by policy, the technician would deem the result a "pass." Such data would be aggregated into pass/fail shelf life trend reports, and the quality manager would assume that the data are indicative of acceptable conditions in the production environment and within the product. However, what if the analytical unit sampled at T14 were to be instead sampled at T28 or T60? Would it have still had acceptable plate counts or would that population have grown much more rapidly than within other samples? Is this shelf life test based upon destructive sampling of only four analytical units enough from which to draw conclusions, or does it only provide misleading, anecdotal information? Table 1.3 demonstrates how inclusion of duplicate samples might avoid instances of misinterpretation of microbial counts in shelf life testing. Table 1.2 data represent a scenario where a sample set of four analytical units were collected the beginning of the production run, while a second sample set of four analytical units were collected from near the end of the production run. T0 samples from both sets were analyzed, and then the remaining units were stored at 4.4 °C. While data generated from the first sample set (Sample 1) show acceptable results, the data from Sample Set 2, which were collected from the production line later, surpassed the acceptable limits for APC at T28. Incidentally, the Lactics data at T28 were still barely acceptable, while APC surpassed the limits. This is a common observation in plant-derived samples, as often the populations on APC are inclusive of many of the same lactic acid bacteria that were enumerated on the Lactics count. At any rate, the result is much different, deemed a "Fail" at T28. The span of time between analyses of samples should factor into interpretation of results. For example, a failure from at least one analytical unit occurred at T28, but the prior sample was analyzed at T14. If data were being generated about the number of days of shelf life achieved, then listing either T14 or T28 would falsely skew the result. An arbitrary number half way between the two (i.e., T21) is a suitable way to estimate days achieved.

Table 1.2 Routine shelf life data from plant-produced samples for which criteria for end of shelf life is 1,000,000 CFU/g for either APC or Lactics

Day of sample	Aerobic plate count (CFU/g)	Lactic acid bacteria count (CFU/g)	Interpretation of result
T0	<1000	1000	Acceptable
T14	322,000	103,000	Acceptable
T28	15,000	24,000	Acceptable
T60	700,000	475,000	Acceptable

Table 1.3 Routine shelf life data from replicate plant produced samples for which criteria for end of shelf life is 1,000,000 CFU/g for either APC or Lactics

Day of sample	Aerobic plate count (CFU/g)		Lactic acid bacteria count (CFU/g)		Interpretation of result
	Sample set 1	Sample set 2	Sample set 1	Sample set 2	
T0	<1000	2000	1000	2000	Acceptable
T14	322,000	250,000	103,000	140,000	Acceptable
T28	15,000	1,075,000	24,000	920,000	Fail
T60	700,000	2,500,000	475,000	1,700,000	Fail

1.2.4 Summary

This book focuses on perishable foods, as there is little to no microbial degradation on low-water activity foods. Indeed, the shelf life of low moisture foods involves principally lipid oxidation (Hu 2016), physical decomposition such as staling, and the resulting organoleptic failure over time. Also, this work is not the same as food spoilage, as that was covered in another title in this series (Sperber and Doyle 2009). Rather, the book focuses on the procedures utilized to establish and monitor shelf life of various food types, with subsequent chapters covering in more detail the metrics for analysis of shelf life of foods, the interpretation of data to set code dates, the types of microorganisms that must be monitored during shelf life testing, and the techniques that can be employed to extend shelf life of perishable foods.

References

Ahumada, Omar, and J. Rene Villalobos. 2009. Application of planning models in the Agri-Food supply chain: A review. *European Journal of Operational Research* 196 (1): 1–20. https://doi.org/10.1016/j.ejor.2008.02.014.

Akkerman, Renzo, Poorya Farahani, and Martin Grunow. 2010. Quality, safety and sustainability in Food distribution: A review of quantitative operations management approaches and challenges. *OR Spectrum* 32 (4): 863–904. https://doi.org/10.1007/s00291-010-0223-2.

Alice, Giusto. 2017. "Language and Food Safety: The 'Zombie Meat' Scandal." *Вестник Санкт-Петербургского Университета. Серия 13. Востоковедение. Африканистика* 9 (1).

Amorim, P., H. Meyr, C. Almeder, and B. Almada-Lobo. 2013. Managing perishability in production-distribution planning: A discussion and review. *Flexible Services and Manufacturing Journal* 25 (3): 389–413. https://doi.org/10.1007/s10696-011-9122-3.

Anonymous. 1996. *"Bill Emerson Good Samaritan Food Donation Act."* 42 U.S. Code § 1791. United States of America.

———. 2011. Food safety modernization act domestic and foreign facility re-inspections, recall, and importer re-inspection user fee rates for fiscal year 2012. *Federal Register* 76 (147): 45820–45825.

Apaiah, Radhika, K. Eligius, M.T. Hendrix, Gerrit Meerdink, and Anita R. Linnemann. 2005. Qualitative methodology for efficient Food chain design. *Trends in Food Science & Technology* 16 (5): 204–214. https://doi.org/10.1016/j.tifs.2004.09.004.

Aschemann-Witzel, Jessica, Ilona de Hooge, Pegah Amani, Tino Bech-Larsen, and Marije Oostindjer. 2015. Consumer-related Food waste: Causes and potential for action. *Sustainability* 7 (6): 6457–6477. https://doi.org/10.3390/su7066457.

Bánáti, Diána. 2011. Consumer response to Food scandals and scares. *Trends in Food Science & Technology* 22 (2): 56–60.

Benson, Andrew, K. Jairus, R.D. David, Stefanie Evans Gilbreth, Gordon Smith, Joseph Nietfeldt, Ryan Legge, Jaehyoung Kim, et al. 2014. Microbial successions are associated with changes in chemical profiles of a model refrigerated fresh pork sausage during an 80-day shelf life study. Edited by D W Schaffner. *Applied and Environmental Microbiology* 80 (17): 5178–5194. https://doi.org/10.1128/AEM.00774-14.

Brul, S., and P. Coote. 1999. Preservative agents in foods: Mode of action and microbial resistance mechanisms. *International Journal of Food Microbiology* 50 (1–2): 1–17. https://doi.org/10.1016/S0168-1605(99)00072-0.

Cauchie, Emilie, Mathieu Gand, Gilles Kergourlay, Bernard Taminiau, Laurent Delhalle, Nicolas Korsak, and Georges Daube. 2017. The use of 16S RRNA gene Metagenetic monitoring of refrigerated Food products for understanding the kinetics of microbial subpopulations at different storage temperatures: The example of white pudding. *International Journal of Food Microbiology* 247: 70–78. https://doi.org/10.1016/j.ijfoodmicro.2016.10.012.

Dalgaard, Paw, Ole Mejlholm, and Hans Henrik Huss. 1997. Application of an iterative approach for development of a microbial model predicting the shelf-life of packed fish. *International Journal of Food Microbiology* 38 (2–3): 169–179. https://doi.org/10.1016/S0168-1605(97)00101-3.

David, Jairus R.D., Larry R. Steenson, and P. Michael Davidson. 2013. Expectations and applications of natural antimicrobials to foods. *Food Protection Trends* 33 (4): 238–247. http://www.researchgate.net/profile/Jairus_David/publication/270273839_PEER-REVIEWED_ARTICLE/links/54a4239e0cf257a636071d3b.pdf.

de Jonge, Janneke, Hans Van Trijp, Reint Jan Renes, and Lynn J. Frewer. 2010. Consumer confidence in the safety of food and newspaper coverage of food safety issues: A longitudinal perspective. *Risk Analysis* 30 (1): 125–142. https://doi.org/10.1111/j.1539-6924.2009.01320.x.

Fu, Bin, and Theodore P. Labuza. 1993. Shelf-life prediction: Theory and application. *Food Control* 4: 125–133. https://doi.org/10.1016/0956-7135(93)90298-3.

Gámbaro, Adriana, S. Fiszman, A. Giménez, P. Varela, and A. Salvador. 2004. Consumer acceptability compared with sensory and instrumental measures of white Pan bread: Sensory shelf-life estimation by survival analysis. *Journal of Food Science* 69 (9): S401–S405.

Geng, Shu, Xu Liu, and Roger Beachy. 2015. New Food safety law of China and the special issue on Food safety in China. *Journal of Integrative Agriculture* 14 (11): 2136–2141. https://doi.org/10.1016/S2095-3119(15)61164-9.

Gram, Lone, Lars Ravn, Maria Rasch, Jesper Bartholin Bruhn, Allan B. Christensen, and Michael Givskov. 2002. Food spoilage-interactions between Food spoilage Bacteria. *International Journal of Food Microbiology* 78 (1–2): 79–97. https://doi.org/10.1016/S0168-1605(02)00233-7.

Grunert, Klaus G. 2002. Current issues in the understanding of consumer Food choice. *Trends in Food Science & Technology* 13 (8): 275–285. https://doi.org/10.1016/S0924-2244(02)00137-1.

Grunow, Martin, and Selwyn Piramuthu. 2013. RFID in highly perishable Food supply chains - remaining shelf life to supplant expiry date? *International Journal of Production Economics* 146 (2): 717–727. https://doi.org/10.1016/j.ijpe.2013.08.028.

Gunders, Dana. 2012. Wasted: How America is losing up to 40 percent of its Food from farm to fork to landfill. *Natural Resources Defense Council*: 1–26.

Hough, G., K. Langohr, G. Gómez, and A. Uria. 2003. Survival analysis applied to sensory shelf life of foods. *Journal of Food Science* 68 (1): 359–362. https://doi.org/10.1111/j.1365-2621.2003.tb14165.x.

Hu, Min. 2016. Oxidative stability and shelf life of low-moisture foods. In *Oxidative stability and shelf life of foods containing oils and fats*, ed. Min Hu and Charlotte Jacobson, 313–371. Amsterdam: AOCS Press.

in't Veld, Jos H. Huis. 1996. Microbial and biochemical spoilage of foods: An overview. *International Journal of Food Microbiology* 33 (1): 1–18. https://doi.org/10.1016/0168-1605(96)01139-7.

Joint Industry Unsaleables Leadership Team. 2008. *"Joint Industry Unsaleables Report: The Real Causes and Actionable Solutions."*

Kaipia, Riikka, Iskra Dukovska-Popovska, and Lauri Loikkanen. 2013. Creating sustainable fresh Food supply chains through waste reduction. *International Journal of Physical Distribution and Logistics Management* 43 (3): 262–276. https://doi.org/10.1108/IJPDLM-11-2011-0200.

Kilcast, David. 2001. Shelf-life evaluation of foods (second edition). *International Journal of Food Science & Technology* 36 (8): 856. https://doi.org/10.1046/j.1365-2621.2001.0530b.x.

Koutsoumanis, Konstantinos. 2001. Predictive modeling of the shelf life of fish under nonisothermal conditions. *Applied and Environmental Microbiology* 67 (4): 1821–1829. https://doi.org/10.1128/AEM.67.4.1821-1829.2001.

Kreyenschmidt, J., A. Hübner, E. Beierle, L. Chonsch, A. Scherer, and B. Petersen. 2010. Determination of the shelf life of sliced cooked ham based on the growth of lactic acid Bacteria in different steps of the chain. *Journal of Applied Microbiology* 108 (2): 510–520. https://doi.org/10.1111/j.1365-2672.2009.04451.x.

Labuza, T.P. 1984. Application of chemical kinetics to deterioration of foods. *Journal of Chemical Education* 61 (4): 348. https://doi.org/10.1021/ed061p348.

Lebersorger, S., and F. Schneider. 2014. Food loss rates at the Food retail, influencing factors and reasons as a basis for waste prevention measures. *Waste Management (New York, NY)* 34 (11): 1911–1919.

Man, C.M.D. 2004. Shelf-life testing. In *Understanding and Measuring the Shelf-Life of Food*, 340–356. Sawston: Woodhead Publishing. https://doi.org/10.1016/B978-1-85573-732-7.50019-6.

Man, C.M.D., and Adrian A. Jones. 1994. *Shelf life evaluation of foods.* New York: Springer.

Martins, R.C., V.V. Lopes, A.A. Vicente, and J.A. Teixeira. 2008. Computational shelf-life dating: Complex systems approaches to Food quality and safety. *Food and Bioprocess Technology* 1 (3): 207–222. https://doi.org/10.1007/s11947-008-0071-0.

McMeekin, Thomas A., and Thomas Ross. 1996. Shelf life prediction: Status and future possibilities. *International Journal of Food Microbiology* 33: 65–83. https://doi.org/10.1016/0168-1605(96)01138-5.

Mena, Carlos, B. Adenso-Diaz, and Oznur Yurt. 2011. The causes of Food waste in the supplier–retailer Interface: Evidences from the UK and Spain. *Resources, Conservation and Recycling* 55 (6): 648–658.

National Advisory Committee on Microbiological Criteria for Foods. 2010. Parameters for determining inoculated pack/challenge study protocols. *Journal of Food Protection* 73: 140–202. https://doi.org/10.4315/0362-028X-73.1.140.

Newsome, Rosetta, Chris G. Balestrini, Mitzi D. Baum, Joseph Corby, William Fisher, Kaarin Goodburn, Theodore P. Labuza, Gale Prince, Hilary S. Thesmar, and Frank Yiannas. 2014. Applications and perceptions of date labeling of Food. *Comprehensive Reviews in Food Science and Food Safety* 13 (4): 745–769.

Office of Technology Assessment, U.S. Government Printing Office. 1979. *"Open Shelf-Life Dating of Food."*

Pothakos, Vasileios, Bernard Taminiau, Geert Huys, Carine Nezer, Georges Daube, and Frank Devlieghere. 2014. Psychrotrophic lactic acid Bacteria associated with production batch recalls and sporadic cases of early spoilage in Belgium between 2010 and 2014. *International Journal of Food Microbiology* 191: 157–163. https://doi.org/10.1016/j.ijfoodmicro.2014.09.013.

Remenant, Benoît, Emmanuel Jaffrès, Xavier Dousset, Marie France Pilet, and Monique Zagorec. 2015. Bacterial spoilers of food: Behavior, fitness and functional properties. *Food Microbiology* 45: 45–53. https://doi.org/10.1016/j.fm.2014.03.009.

Roberts, T.A., and R.B. Tompkin. 1996. *Microorganisms in foods 5: Characteristics of microbial pathogens.* Vol. 5. New York: Springer Science & Business Media.

Rouger, Amélie, Nicolas Moriceau, Hervé Prévost, Benoît Remenant, and Monique Zagorec. 2018. Diversity of bacterial communities in French chicken cuts stored under modified atmosphere packaging. *Food Microbiology* 70: 7–16. https://doi.org/10.1016/j.fm.2017.08.013.

Singh, R.P. 1994. Scientific principles of shelf life evaluation. In *Shelf Life Evaluation of Foods*, 3–26. Cham: Springer Nature. https://doi.org/10.1007/978-1-4615-2095-5_1.

Singh, T.K., and K.R. Cadwallader. 2004. Ways of measuring shelf-life and spoilage. In *Understanding and Measuring the Shelf-Life of Food*, 165–183. Sawston: Woodhead Publishing. https://doi.org/10.1016/B978-1-85573-732-7.50013-5.

Soliva-Fortuny, Robert C., and Olga Martín-Belloso. 2003. New advances in extending the shelf-life of fresh-cut fruits: A review. *Trends in Food Science & Technology* 14 (9): 341–353.

Sperber, William H., and Michael P. Doyle. 2009. *Compendium of the microbiological spoilage of foods and beverages*. New York: Springer.

Tian, Feng. 2016. An agri-food supply chain traceability system for China based on RFID & Blockchain technology. In *2016 13th international conference on service systems and service management (ICSSSM)*, 1–6. https://doi.org/10.1109/ICSSSM.2016.7538424.

———. 2017. "A supply chain traceability system for food safety based on HACCP, blockchain & internet of things." *14th International Conference on Services Systems and Services Management, ICSSSM 2017-Proceedings*. https://doi.org/10.1109/ICSSSM.2017.7996119.

Tsiros, Michael, and Carrie M. Heilman. 2005. The effect of expiration dates and perceived risk on purchasing behavior in grocery store perishable categories. *Journal of Marketing* 69 (2): 114–129. https://doi.org/10.1509/jmkg.69.2.114.60762.

U.S. Department of Agriculture, and Food Safety and Inspection Service. 2015. "*Kraft Heinz Foods Company Recalls Turkey Bacon Products Due To Possible Adulteration.*" https://www.fsis.usda.gov/wps/portal/fsis/topics/recalls-and-public-health-alerts/recall-case-archive/archive/2015/recall-113-2015-release.

———. 2015. "*Control of Listeria Monocytogenes in ready-to-eat meat and poultry products.*" Vol. 80.

U.S. Department of Justice, and Office of Public Affairs. 2015. "*Quality Egg, Company Owner and Top Executive Sentenced in Connection with Distribution of Adulterated Eggs.*" https://www.justice.gov/opa/pr/quality-egg-company-owner-and-top-executive-sentenced-connection-distribution-adulterated.

U.S. Food and Drug Administration. 2015. "*Current Good Manufacturing Practice, Hazard Analysis, and Risk-Based Preventive Controls for Human Food.*" 21 C.F.R. § 117.

———. 2017. "*FDA Food Code.*" https://doi.org/10.1016/j.parint.2011.08.011.

Van Boekel, M.A.J.S. 2008. Kinetic modeling of Food quality: A critical review. *Comprehensive Reviews in Food Science and Food Safety* 7: 144–158. https://doi.org/10.1111/j.1541-4337.2007.00036.x.

Van Boxstael, S., F. Devlieghere, D. Berkvens, A. Vermeulen, and M. Uyttendaele. 2014. Understanding and attitude regarding the shelf life labels and dates on pre-packed Food products by Belgian consumers. *Food Control* 37 (1): 85–92. https://doi.org/10.1016/j.foodcont.2013.08.043.

Wee, H.M., 1993. Economic production lot size model for deteriorating items with partial back-ordering. Computers & Industrial Engineering, 24(3), pp. 449–458.

Wansink, Brian, and Alan O. Wright. 2006. 'Best if used by …' how freshness dating influences Food acceptance. *Journal of Food Science* 71 (4): S354–S357. https://doi.org/10.1111/j.1750-3841.2006.00011.x.

Xu, Y. and Sarker, B.R., 2003. Models for a family of products with shelf life, and production and shortage costs in emerging markets. Computers & Operations Research, 30(6), pp. 925–938.

Chapter 2
Food Safety Factors Determining Shelf Life

Margaret D. Hardin

2.1 Introduction

The shelf life of food products is often defined as the recommended maximum amount of time that food products can be stored under specified conditions of temperature, humidity, and other external factors while maintaining acceptable quality without exhibiting spoilage. Spoilage of perishable foods may be the result of changes in the sensory characteristics of the product, such as the development of off-flavors, off-colors, gas, or slime, which make the product undesirable or unacceptable to the consumer, or by exceeding a certain level of indicator organisms specified for the product. Many of these undesirable changes are the result of microbial growth and the associated metabolic activity and by-products of the organism(s).

Food products naturally contain microorganisms that can include spoilage and/or pathogenic microorganisms. However, the growth of pathogenic microorganisms in food products seldom coincides with noticeable spoilage characteristics such as off-odor, discoloration, gas, or slime. Microorganisms that spoil foods and those that are of public health significance can do so at various times in the process including before and during preparation and/or processing, under normal conditions of storage and intended use, particularly if not destroyed or controlled by normal processing techniques. While a range of methods such as salting, curing, smoking, freezing, and canning have been successfully used over the years for extending the shelf life of foods, consumers seem to prefer fresh foods over frozen or shelf-stable foods. In addition, consumers are becoming more aware of product labels and are demanding products that are preservative-free and have minimal processing. Consumers are

M. D. Hardin (✉)
Vice President of Technical Services, IEH Laboratories and Consulting Group,
Lake Forest Park, WA, USA
e-mail: mh@iehinc.com

© Springer Nature Switzerland AG 2021
P. J. Taormina, M. D. Hardin (eds.), *Food Safety and Quality-Based Shelf Life of Perishable Foods*, Food Microbiology and Food Safety,
https://doi.org/10.1007/978-3-030-54375-4_2

avoiding products containing ingredients that are not recognizable or perceived as natural. Consumer demand for so-called clean label products is increasing. These ready-to-eat (RTE) and perishable fresh foods that have an enhanced but limited shelf life rely upon time and temperature as critical factors for maintaining microbiological quality and safety.

With few exceptions, refrigerated foods have had a very good history of safety. However, over the past 20–30 years, innovations in refrigerated foods with an extended shelf life have gained in popularity. Various methods of processing (e.g., flash pasteurization, high pressure processing) and packaging (e.g., vacuum packaging and modified atmosphere packaging) have added extra days, and even weeks or months, to the refrigerated shelf life of these products. The benefits of an extended shelf life include products that can remain on the shelf at retail or in a consumer's refrigerator for longer periods of time; reduced waste; reduced product returned to the manufacturer; increased distribution of perishable products over a wider geographical area; and the production and extended storage of seasonal products. These refrigerated foods include conventional RTE products such as luncheon meats and sausages (both cured and uncured), as well as refrigerated salads containing meat, egg and/or seafood, fresh pasta and pasta sauces, soups, sauces, entrees, complete meals, as well as minimally processed and novel products such as sous-vide type foods (Marth 1998; Austin 2001).

2.2 Food Safety Concerns for Extended Shelf Life Foods

With the growing demand for minimally processed refrigerated foods with extended shelf life, concern has increased related to the risk associated with RTE foods that require refrigeration and support the growth of psychrotrophic pathogens (NACMCF 2005; Marth 1998; FDA 2003). Four pathogens of concern capable of growth at refrigeration temperatures have been identified for this group of products (i.e., refrigerated RTE foods): *Yersinia enterocolitica, Bacillus cereus,* nonproteolytic *Clostridium botulinum*, and *Listeria monocytogenes*. These four organisms were identified by an Institute of Food Technologists (IFT) expert panel on food safety and nutrition in 1998 (Marth 1998) and again in 2005 by US National Advisory Committee on Microbiological Criteria in Foods (NACMCF), when they performed a hazard analysis to identify microorganism of concern when establishing safety-based date labeling (NACMCF 2005). While all of these pathogens are capable of growth at refrigeration temperatures, many are underreported due to the self-limiting nature of the disease, lack of routine laboratory testing, limited availability of rapid methods, and/or lack of culture procedures available for isolating the organisms. In addition, some of the illnesses, such as those associated with *B. cereus*, are not reportable diseases and are therefore likely underestimated in official reporting systems.

2.2.1   Yersinia enterocolitica

The genus *Yersinia* includes both pathogenic and nonpathogenic strains of the organism. Of the pathogenic strains, *Yersinia enterocolitica* is the strain most commonly associated with outbreaks of foodborne illness. *Y. enterocolitica* is generally ubiquitous in the environment and has been recovered from a wide variety of animals (dogs, cats, birds, monkeys, and shellfish), foods (milk and raw milk, raw pork, prepared foods, vegetables, etc.) and water (Barton and Robins-Browne 2003; Austin 2001; ICMSF 1996; Kapperud 1991; Robins-Browne 2001; Ackers et al. 2000). Low recovery rates are often attributable to a lack of routine testing in clinical situations and during outbreak investigations and to the limited of sensitivity of available culture methods (ICMSF 1996; Robins-Browne 2001; Barton and Robins-Browne 2003). Although pets may be occasional carriers of *Y. enterocolitica*, pigs are considered a major primary source of yersiniosis infections in humans particularly those caused by bioserotype 4,O:3. Symptom of the illness, yersiniosis, include abdominal pain (sometimes confused with appendicitis), headache, fever, diarrhea, nausea, vomiting. Data has shown that *Y. enterocolitica* is able to grow over a wide range of temperatures from temperatures near 0 °C to 42 °C, particularly when conditions for growth are most favorable (ICMSF 1996; Kapperud 1991). The ability of the organism to grow in foods when stored at very low temperatures will vary with other factors including substrate, pH, gaseous atmosphere, salt, preservatives, or competing flora (ICMSF 1996; Kapperud 1991; Barton and Robins-Browne 2003).

The true incidence of foodborne yersiniosis is uncertain for various reasons: few outbreaks of foodborne illness are investigated; yersiniosis has only recently been known to be food or water borne; long periods of time may be required using cold enrichment to recover certain strains from food; and not all clinical laboratories routinely test for the organism. In addition, there are limited methods available that differentiate between pathogenic and nonpathogenic strains and species of the organism, and serological and biochemical testing is outside of the scope for most laboratories. Advances in molecular methods will likely improve detection of *Y. enterocolitica* in foods. Although outbreaks of illness are uncommon, the foods generally associated with illness outbreaks of yersiniosis include milk, water, produce, and undercooked pork products. Some person-to-person transmission of the disease has also been reported (Todd et al. 2007). Pigs appear to be the major source of *Y. enterocolitica* in foods. Some milk borne outbreaks have been reported in pasteurized milk; however, cross-contamination from milk crates used on pig farms was identified as the source (Barton and Robins-Browne 2003). During an outbreak investigation of illnesses due to contaminated tofu in Washington State, the same strain of *Y. enterocolitica* serotype O:8 was isolated from both the tofu and the plant's untreated spring water (Tacket et al. 1985). Untreated water has been identified as a source of infection in other cases. Spring water was identified as the source in a case of *Yersinia enterocolitica* septicemia in New York State (Keet 1974) and untreated well water with a small outbreak of gastroenteritis in Canada (Thompson and Gravel 1986). A more recent outbreak of yersiniosis in Norway in 2014 identified salad mix containing imported radicchio rosso as a source of the illnesses (MacDonald et al. 2016). Although the organism was never isolated from the salad mix, numerous failures in hygiene were observed at the production facility, including infrequent changing of the water used to rinse the produce. The longer shelf life of the product was also cited as a contributing factor. This was not the first outbreak associated with produce in Norway. In 2011, an outbreak of yersiniosis was associated with ingestion of ready-to-eat salad mix (MacDonald et al. 2012). In this case, radicchio rosso was identified as the likely source of infection due to the fact that it can be stored for several months and was the only ingredient included in the suspected salad mix that had delivery, production, and storage dates consistent with the outbreak. Although investigators were unable to conclusively link any of the isolates from the salad ingredients to the human *Y. enterocolitica* isolates, they concluded that finding nonpathogenic *Yersinia* in packaged salads reinforced that the environment of the food product was processed in was conducive to the persistence of the bacterium.

As previously mentioned, pigs are a major source of the organism. Outbreaks of yersiniosis, in the US, primarily with infants and children, have been associated with cross-contamination during the preparation of raw pork, specifically pork chitterlings (Abdel-Hag et al. 2000; Lee et al. 1990; MMWR 2003). Traditionally prepared chitterlings are thoroughly cooked and infections have not been associated with the consumption of the final product; however, preparation of chitterlings involves a lengthy and wet process of cleaning and cooking large amounts of raw pork intestines that may contain the organism. Therefore, during preparation of the product, there is considerable opportunity for cross-contamination to occur, and outbreaks have been reported to other prepared foods, people, baby bottles, and toys, with infants and young children being particularly susceptible to the disease (Abdel-Hag et al. 2000; Lee et al. 1990). Educational materials targeting consumers have been developed focusing on preparation of chitterlings in the home (USDA FSIS 2011). While education of the public on preparation of this product has had some success, cases do still occur. Since *Y. enterocolitica* is capable of multiplying a very low temperatures, refrigerated storage is not generally recommended a means of preventing outbreaks; however, refrigeration temperatures do prolong lag periods (ICMSF 1996). Prevention of cross-contamination from untreated water and raw foods to cooked foods such as unpasteurized milk and raw pork is an essential aspect of a control program (ICMSF 1996; Barton and Robins-Browne 2003; USDA FSIS 2011).

2.2.2   Bacillus cereus

Bacillus cereus is a spore-forming bacterium that has been isolated from a wide variety of products. Due to its widespread presence in nature, it is virtually impossible to obtain raw product that are free of *B. cereus* spores. Milk products and products of plant origin are the main sources of *B. cereus* (Granum 2001; Jenson and Moir 2003). The bacterium may be transferred to other food products and may survive, as spores, in heated products where competition for other bacteria is not usually present. The growth and survival of *B. cereus* have been studied extensively over a range of temperatures, pH values, salt concentrations, preservatives, and other factors. Some strains are psychrotrophic being able to grow at 4–5 °C. Psychrotrophic strains have been a problem for the dairy industry particularly in certain countries and for situations where maintaining low temperatures (<7 °C) cannot always be assured (Notermans et al. 1997; Granum 2001).

   B. cereus is the cause of two different types of food poisoning: an emetic type and a diarrheal type. The emetic syndrome is caused by the ingestion of the heat-stable emetic toxin produced in foods, and the diarrheal syndrome is mainly due to the ingestion of *B. cereus* cells in foods followed by toxin production in the small intestine (Granum and Lund 1997; Granum 2001). Carlin et al. (2006) evaluated 100 representative strains of *B. cereus* selected from a total collection of 372 *B. cereus* strains in order to investigate differences in the growth limits and heat-resistance profiles of emetic toxin-producing and non-emetic toxin-producing strains of *B. cereus*. Emetic toxin-producing strains were able to grow at 48 °C; however, none of the emetic toxin-producing strains were able to grow below 10 °C and spores from the emetic toxin-producing strains showed a higher heat resistance at 90 °C and a lower germination, particularly at 7 °C, than spores from the other strains. The authors concluded that while emetic toxin-producing strains of *B. cereus* pose a particular risk in heat-processed or preheated foods that are kept warm (such as hot-holding in restaurants), they will not pose a risk in refrigerated foods (Carlin et al. 2006). Of the non-emetic toxin-producing strains, 50 strains (28 diarrheal and 28 food-environmental strains) out of 83 were able to grown at 4 °C or 7 °C. All of the diarrheal strains showed lower D-values that emetic toxin-producing strains, and none of the diarrheal strains produced the emetic toxin (Carlin et al. 2006). Granum (1994) evaluated different strains of enterotoxigenic *B. cereus* and the inactivation of toxin following exposure to the low pH and proteolytic enzymes of the stomach. The researcher concluded that it is likely that in the case of the diarrheal illness, food poisoning is caused by ingestion of cells or spores rather than by pre-formed enterotoxin. In addition, the level of enterotoxin produced by different strains of *B. cereus* varies, making it possible that only a few of the enterotoxigenic strains are of public health significance. The author also concluded that ingestion of 10^4 to 10^7 cells or spores are the main cause of food poisoning associated with this illness syndrome.

   Outbreaks of foodborne illness have generally involved food that has been heat treated and growth of surviving spores, in absence of any competitive flora, are the source of the illness. *B. cereus* has frequently been isolated from a wide variety of foods and in addition to products of plant origin such as rice, pasta and spices, dairy products, including both raw and pasteurized milk, are the most common food vehicles for *B. cereus* (Granum 2001; Granum et al. 1993). Although the actual number of foodborne illness outbreaks associated with *B. cereus* is likely underestimated, a few outbreaks of foodborne illness have been reported from psychrotrophic, enterotoxin-producing strains of *B. cereus*. Psychrotrophic strains of *B. cereus* were recovered during outbreak investigations that occurred in Spain and The Netherlands between 1986 and 1989. The food poisoning strains were recovered from various dairy products including pasteurized milk and grew within a temperature range of between 4 and 37 °C (van Netten et al. 1990). Spores of *B. cereus* in raw milk have been reported to survive pasteurization and subsequently colonize production equipment. Lin et al. (1998) isolated both spores and vegetative cells of *B. cereus* from raw milk, pasteurized milk, and environmental samples. Most of the isolates obtained from the pasteurized milk and final products belonged to the same sub-groups as the strains germinated from spores in raw milk suggesting that spores in raw milk were a major source in pasteurized milk. Strains of *B. cereus* isolated from environmental samples came from HTST pipes, pasteurized milk tanks, and fillers, suggesting that they may be a potential reservoir for *B. cereus* within the facility and a potential contributing factor post-pasteurization contamination with the organism.

Since *B. cereus* is a spore-former and is ubiquitous in the environment, a low level of contamination can be expected in most foods. However, low numbers of *B. cereus* vegetative cells and/or its spores are not expected to cause problems unless growth is permitted to occur. Foods implicated in cases of *B. cereus* associated illness usually contain at a range of 10^5 to 10^{7-8} viable cells or spores due to differences in the amounts of enterotoxin produced by different strains (Granum 2001; Granum 1994). Control in foods relies on complete destruction by heating or other lethality treatment designed to control germination of spores or prevent multiplication to hazardous levels in foods. While psychrotrophic strains are capable of growth during shelf life and have the potential to cause illness, few outbreaks have been reported in refrigerated foods held under proper refrigeration. In addition, the relatively short duration and mild nature of the illnesses caused by *B. cereus* have kept this organism in the epidemiological background, particularly when compared to other more prominent foodborne infections and intoxications (ICMSF 1996; Granum 2001; NACMCF 2005).

2.2.3 *Non-Proteolytic* **Clostridium botulinum**

Clostridium botulinum is an anaerobic, spore-forming microorganism capable of producing one or more biological neurotoxins (types A–H). Spores of *C. botulinum* are widely distributed in nature and are commonly found in soil, sediments, and water. The species is divided up into four groups (I–IV) based on DNA homology and physiological differences. Most outbreaks of human botulism are caused by Group I (proteolytic) strains which produce toxins of type A, B, or F and Group II (non-proteolytic) strains which produce type B, E, or F. The optimal temperature for growth of Group I proteolytic strains is between 35 °C and 40 °C with a minimum growth temperature of 10 °C (ICMSF 1996; Austin 2001). Therefore, control of growth of Group I strains can be achieved by storage of food products at temperatures below 10 °C. Outbreaks of foodborne illness associated with Group I strains have often involved incorrectly canned or retorted foods.

Non-proteolytic Group II strains of *C. botulinum* have an optimum temperature for growth of 28–30 °C and will grow at temperatures as low as 3 °C (Graham et al. 1997). Growth is inhibited by 5% salt and a water activity of 0.97 in a NaCl solution (Graham et al. 1997; Austin 2001). Outbreaks associated with non-proteolytic *C. botulinum* often involve smoked, dried, or salted fish, and type E neurotoxin (Peck 2006; EFSA 2016). Most recently, an outbreak of *C. botulinum* type E associated with dried and salted fish was reported in Germany and Spain, however, and as is the case with many reported outbreaks, no details were provided regarding packaging and storage conditions at the time of the reported illnesses (EFSA 2016). While outbreaks of illness associated with Group II strains occur infrequently, the severity of the illness associated with the neurotoxin produced by this organism makes any outbreaks of botulism too many.

Non-proteolytic Group II strains are of particular concern for refrigerated products with an extended shelf life particularly when stored in MAP or vacuum packaging. As the shelf life of refrigerated foods packaged in reduced oxygen packaging (ROP) is increased, more time is available for *C. botulinum* growth and toxin formation. As mentioned previously, pathogenic psychrotrophic microorganisms do not tend to impart a sensory defect that is perceived as spoilage. Group II *C. botulinum* strains are *non-proteolytic*, and therefore, no off-odor or evidence of spoilage may coincide with toxin development. Although storage at temperatures below 3 °C would prevent the growth of the organism, these temperatures are difficult to achieve on a consistent basis, if at all, particularly during distribution, storage, and display at retail or in a home refrigerator (Lund and Peck 1994; ICMSF 1996; Audits International 1999; USDA FSIS 2010; USDA CFSAN 2003). As storage temperatures increase, the time required for toxin formation is significantly shortened, and temperature abuse (>10 °C) could allow Group I strains to grow as well (Szabo and Gibson 2003). Group II *C. botulinum* have also been identified as a particular risk for minimally processed heated and chilled sous-vide products. These products are cooked under vacuum in sealed pouches (oxygen barrier bags), at precise (and sometimes low) temperatures, and often for long times. Sous-vide can be used to prepare foods with an extended shelf life for retail sale or use in food service. For some products, this results in minimally processed foods that may be undercooked depending on recipe. The lack of heat treatment sufficient to destroy spores of *C. botulinum* and sometimes lack of thorough heating prior to consumption, combined with packaging in an atmosphere with reduced or no oxygen, increases the risk of botulism from these foods. This places the reliance for the safety of these products solely on adequate refrigeration and time. However, as was previously mentioned above, the temperatures required to preclude the growth of, and toxin production by, non-proteolytic *C. botulinum* are not necessarily achieved in the distribution, retailing, and storage of chilled foods.

Modified atmosphere packaging, once considered a new technology, is continually being applied to new and novel foods and food processes. The main purpose of reduced oxygen/modified atmosphere packaging for processed foods is to inhibit the growth of aerobic spoilage microbes and spoilage characteristics associated with these types of proteolytic and lipolytic spoilage organisms, thereby increasing the shelf life of the product. However, reducing the levels of oxygen in a product and inhibiting the growth of competing flora allow for the outgrowth of *C. botulism* spores if temperature is not adequately controlled, particularly during extended shelf life. Control measures for specific categories of products including meat and meat products, fishery products, fruits and vegetables, dairy products, and refrigerated processed foods of extended shelf life with reduced oxygen packaging, also known as refrigerated processed foods of extended durability (REPFEDs), have been outlined in numerous publications (FSA 2017; NZFA 2005; Szabo and Gibson 2003). Although the minimum growth temperature for non-proteolytic strains of *C. botulinum* recommended in most of these publications is 3.3 °C, Graham et al. (1997) reported growth and toxin formation of non-proteolytic strains of *C. botulinum* in 5–6 weeks. Due to the risk for growth of Group II strains of *C. botulinum* at refrigeration temperatures, several regulatory agencies and advisory committees have that for prepared chilled foods with extended shelf life (>10 days), additional controls such as pH, water activity, % salt, 6D heat treatment, antimicrobial additives, should also be used singly or in combination with temperature to preclude growth and toxin production of psychrotrophic *C. botulinum* (FSA 2017; NZFA 2005; Szabo and Gibson 2003).

2.2.4 Listeria monocytogenes

Listeria is a ubiquitous organism, widely distributed in the environment including soil, vegetation, silage, feces, sewage, and water. The genus *Listeria* contains several species with *Listeria monocytogenes* being the one associated with human illness outbreaks. *Listeria monocytogenes* was first identified in 1923 and associated, at the time, with zoonotic diseases including abortions in cattle and sheep. *L. monocytogenes* was first recognized as a significant cause of human illness in 1981. Since that time, both humans and animals have been recognized as asymptomatic carriers of the organism. Clinical symptoms of the illness range, in healthy adults, from mild flu-like symptoms of noninvasive gastroenteritis to more severe invasive cases of the disease. In the more severe cases, symptoms may include septicemia and meningitis. Infections may become life threatening particularly for individuals with suppressed immune systems such as pregnant women, newborns, elderly, and immunocompromised individuals such as HIV-positive patients, individuals receiving dialysis or chemotherapy. The high mortality rate of between 25 and 35% is a major concern with listeriosis (Bean and Griffin 1990).

Lower growth limits for *L. monocytogenes* have been reported in sterile foods (chicken broth and UHT milk) as low as 0 °C; however, growth at this temperature is slow (Walker et al. 1990). *L. monocytogenes* also grows well under varying gaseous atmosphere aerobic, anaerobic, and microaerophilic conditions (Ingham et al. 1990). *L. monocytogenes* is not resistant to heat, and consequently, thorough cooking will destroy the organism. Food safety concerns and public health implications related to *Listeria monocytogenes* stem primarily from contamination of ready-to-eat food products that receive no additional cooking by the food preparer or consumer. Although outbreaks of foodborne illness have been reported in a wide variety of RTE foods, some foods have been identified as posing a higher risk for *L. monocytogenes*. This includes RTE foods that have the potential to be contaminated with *L. monocytogenes* and which support the growth of the organism to high numbers, particularly due to storage for extended periods of time at refrigeration temperatures (ILSI 2005; FDA 2003; USDA FSIS 2010). Some of the categories of foods that have been identified to be high include unpasteurized milk products, smoked seafood, deli meats, and frankfurters (FDA 2003; USDA FSIS 2010). Differences in risk for *L. monocytogenes* have been reported for products prepared at a commercial production facility versus a retail delicatessen (Gombas et al. 2003; Pouillot et al. 2015; Pradhan et al. 2011). Higher prevalence rates and levels of *L. monocytogenes* have been reported for RTE deli meats and salads packaged at retail as compared to product packaged at a processing facility (Gombas et al. 2003). However, a more recent survey found no significant differences between deli packaged and prepackaged seafood salads and deli-type salads without meat (Luchansky et al. 2017). This multiyear Market Basket Survey also reported that the occurrence of *L. monocytogenes* in the RTE foods tested had decreased and that the prevalence was lower than reported in the Gombas et al. study conducted in 2003.

While it is nearly impossible to eliminate *L. monocytogenes* from raw foods, most measures to control *L. monocytogenes* in RTE products involve continual management and a multiple-hurdle approach: Control measures focusing on the hygienic design of the facility and equipment; good manufacturing practices (GMPs); effective sanitation to avoid cross-contamination of RTE processed products; a stringent environmental sampling program; and when possible, prevention of growth of the organism in the product through the use of antimicrobial ingredients and/or processes, and temperature control. In the environment, limiting product contamination and proper sanitation including employee hygiene, controlling traffic, preventing establishment of the organism in coolers, cold rooms, air handling units, on equipment, and conveying systems of post-lethality areas (ready-to-eat areas) of a processing plant or in a food preparation kitchen in conjunction with a focused and targeted environmental testing program will reduce the risk of product contamination with this organism (Tompkin 2002; USDA 2010; FDA 2003; Carpentier and Cerf 2011).

2.3   Challenge Studies and Shelf Life

While some publications refer to shelf life studies as challenge studies (NACMCF 2010), there are distinct differences between shelf testing and challenge studies. Shelf life studies are an objective means to determine the time a product can be expected to keep under specified storage conditions without appreciable changes in product quality or safety. The end of shelf life is often based on one or more changes in sensory, chemical, functional, physical, and microbiological characteristics of the product. They may be conducted for various reasons such as a new product launch; for a change in product formulation, ingredients, supplier, or packaging; to validate a process; or as part of an on-going QA/QC verification program. Microbiological challenge studies are conducted to determine the ability of a food to support the growth of spoilage organisms or pathogens. Microbiological challenge studies are performed by inoculating a specific level of selected microorganisms into a food product or ingredient to determine to the food safety risk or risk of spoilage. Microbiological challenge studies can be designed and used for a variety of purposes. They can determine if a particular microorganism will grow in a particular food under a specified set of circumstances. Challenge studies may be conducted to evaluate the effectiveness of a lethality step, or of an antimicrobial ingredient or processes, or to simulate what happens to a product during processing, distribution, and/or subsequent handling. The shelf life of a product is often determined during the product development stage prior to any large-scale production or market testing. At this time, the risk for contamination, survival, and growth of pathogenic organisms during shelf life may also be evaluated, particularly for high risk food such as food stored under reduced oxygen for extended periods of time at refrigeration temperatures. When deciding whether a shelf life or challenge study needs to be done, it is important to have the appropriate expertise available, such as an experienced food microbiologist, to help determine the need for the study and assist in the design and in the interpretation of results. An expert food microbiologist will also be able to identify the appropriate target organism, inoculum type and level, and method of inoculation. All known sources of expected and unexpected variability in the product, batch, as well as with the equipment and manufacturing process, need to be taken into consideration. Worst-case scenario conditions should be considered during the study design process. While shelf life trials for refrigerated perishable foods are often conducted at temperatures of 40 °F (4.4 °C) or lower, shelf life trials, particularly those conducted for perishable refrigerated foods, should also include temperatures of mild abuse (45–50 °F/7–10 °C) to reflect actual cold chain temperatures that may occur during commercial distribution and storage or display. These trials should be conducted for several days or weeks beyond the targeted shelf life unless the product fails earlier. Products are then evaluated at set times throughout the trial for physical and chemical changes in the product that may indicate spoilage, as well as conducting microbiological testing for spoilage and/or indicator organisms, and sometimes for pathogens. Several publications are available to provide assistance in designing these types of studies (NACMCF 2010; Scott et al. 2005; Hardin 2012; FSA 2017).

2.4 Summary

As stated in the beginning of this chapter, the development and production of new and novel refrigerated foods with an extended shelf life have gained in popularity over the past 20–30 years.

These precooked, prepackaged convenience foods have increased in popularity for both the foodservice and retail segments of the industry. The types of products that are available are seemingly endless and range from soups and salads to sauces and complete meals and may contain a variety of food ingredients from anywhere in the world. These products are often packaged under vacuum or modified air, have an extended refrigerated shelf life, are not protected by conventional preservation systems such as reduced water activity or pH, and are intended to receive little or no additional heating prior to consumption. This presents a public health concern with respect to the growth of pathogens of concern outlined in this chapter. If present, these organisms can multiply in low-oxygen packaging systems under extend storage at refrigeration temperatures. Shelf life determination of these refrigerated foods must involve an assessment of risk of bacterial pathogens that are capable of growth at refrigeration temperatures. Challenge studies can be used to further elucidate the growth potential of these bacterial pathogens and risk of the product to public health.

References

Abdel-Hag, N.M., B.I. Asmar, W.M. Abuhammour, and W.J. Brown. 2000. *Yersinia enterocolitica* infection in children. *Pediatric Infectious Disease Journal* 19: 954–958.

Ackers, Marta-Louise, Susan Schoenfeld, Markman John, M. Geoffrey Smith, Mabel A. Nicholson, Wallis DeWitt, Daniel N. Cameron, Patricia M. Griffin, and Laurence Slutsker. 2000. An outbreak of *Yersinia enterocolitica* O:8 infections associated with pasteurized milk. *Journal of Infectious Diseases* 181: 1834–1837.

Audits International/FDA. 1999. "*U.S. food temperature evaluation design and summary pages.*" http://foodrisk.org/files/Audits-FDA_temp_study.pdf. Accessed 28 December 2017.

Austin, John W. 2001. Clostridium botulinum. In *Food Microbiology Fundamentals and Frontiers*, ed. Michael P. Doyle, Larry R. Beuchat, and Thomas Montville, 2nd ed., 329–349. Washington, DC: ASM Press.

Barton, Mary D., and Roy M. Robins-Browne. 2003. Yersinia enterocolitica. In *Foodborne Microorganisms of Public Health Significance*, ed. Alisa D. Hocking, 6th ed., 577–596. Marrickville, NSW: Southwood Press.

Bean, N.H., and P.M. Griffin. 1990. Foodborne disease outbreaks in the United States, 1973–1987. Pathogens, vehicles, and trends. *Journal of Food Protection* 53: 804–817.

Carlin, Frédéric, Martina Fricker, Annemarie Pielaat, Simon Heisterkamp, Ranad Shaheen, Mirja Salkinoja Salonen, Birgitta Svensson, Christophe Nguyen-the, and Monika Ehling-Schulz. 2006. Emetic toxin-producing strains of *Bacillus cereus* show distinct characteristics within the *Bacillus cereus* group. *International Journal of Food Microbiology* 109: 132–138.

Carpentier, Brigette, and Olivier Cerf. 2011. Review-persistence of *Listeria monocytogenes* in food industry equipment and premises. *International Journal of Food Microbiology* 145: 1–8.

European Food Safety Authority (EFSA). 2016. Type E botulism associated with fish product consumption—Germany and Spain. *EFSA Technical Report* 2016:EN-1157. http://www.efsa.europa.eu/sites/default/files/1157e.pdf. Accessed 28 December 2017.

Food Standards Agency (FSA). 2017. *The safety and shelf-life of vacuum and modified atmosphere packed chilled foods with respect to non-proteolytic Clostridium botulinum.* United Kingdom: Vacuum Packaging Technical Guidance. https://www.food.gov.uk/sites/default/files/multimedia/pdfs/publication/vacpacguide.pdf. Accessed 28 December 2017.

Gombas, D., Y. Chen, R.S. Clavero, and V.N. Scott. 2003. Survey of *Listeria monocytogenes* in ready-to-eat foods. *Journal of Food Protection* 66: 559–569.

Graham, A.F., D.R. Mason, F.J. Maxwell, and M.W. Peck. 1997. Effect of pH and NaCl on growth from spores of non-proteolytic *Clostridium botulinum* at chill temperature. *Letters in Applied Microbiology* 24: 95–100.

Granum, P.E. 1994. *Bacillus cereus* and its toxins. *Journal of Applied Bacteriology Symposium Supplement* 76: 61S–66S.

Granum, Per Einar. 2001. Bacillus cereus. In *Food Microbiology Fundamentals and Frontiers*, ed. Michael P. Doyle, Larry R. Beuchat, and Thomas Montville, 2nd ed., 373–381. Washington, DC: ASM Press.

Granum, Per Einar, and Terje Lund. 1997. *Bacillus cereus* and its food poisoning toxins. *FEMS Microbiology Letters* 157: 223–228.

Granum, Per Einar, Sigrid Brynestad, and John M. Kramer. 1993. Analysis of enterotoxin production by *Bacillus cereus* from dairy products, food poisoning incidents and non-gastrointestinal infections. *International Journal of Food Microbiology* 17: 269–279.

Hardin, M.D. 2012. Food process validations. In *Microbiological Research and Development for the Food Industry*, ed. Peter J. Taormina, 45–107. Boca Raton. FL: CRC Press.

Ingham, S.C., J.M. Escude, and P. McCowen. 1990. Comparative growth rates of *Listeria monocytogenes* and *Pseudomonas fragi* on cooked chicken loaf stored under air and two modified atmospheres. *Journal of Food Protection* 53: 289–291.

Institute of Life Sciences (ILSI). 2005. Achieving continuous improvement in reductions in foodborne Listeriosis—A risk-based approach. *Journal of Food Protection* 68: 11932–11994.

International Commission on Microbiological Specifications for Food. 1996. *Microorganisms in Foods 5 Microbiological specifications of Food pathogens*. London: Blackie Academic and Professional.

Jenson, Ian, and Catherine J. Moir. 2003. *Bacillus cereus* and other *Bacillus* species. In *Foodborne Microorganisms of Public Health Significance*, ed. Alisa D. Hocking, 6th ed., 445–478. Marrickville, NSW: Southwood Press.

Kapperud, Georg. 1991. *Yersinia enterocolitica* in food hygiene. *International Journal of Food Microbiology* 12: 53–66.

Keet, E.E. 1974. *Yersinia enterocolitica* septicemia. Source of infection and incubation period identified. *New York State Journal of Medicine* 74: 2226–2230.

Lee, Lisa A., A. Russell Gerber, David R. Lonsway, J. David Smith, Geraldine P. Carter, Nancy D. Phur, Christine M. Parrish, R. Keith Sikes, Robert J. Finton, and Robert V. Tauxe. 1990. *Yersinia enterocolitica* O:3 infections in infants and children, associated with the household preparation of chitterlings. *The New England Journal of Medicine* 322: 984–987.

Lin, S., H. Schraft, J.A. Odumeru, and M.W. Griffiths. 1998. Identification of contamination sources of *Bacillus cereus* in pasteurized milk. *International Journal of Food Microbiology* 43: 159–171.

Luchansky, John B., Yuhuan Chen, Anna C.S. Porto-Fett, Régis Pouillot, Bradley A. Shoyer, Rachel Johnson-DeRycke, Denise R. Elben, Karin Hoelzer, William K. Shaw Jr., Jane M. Van Doren, Michelle Catlin, Jeehyun Lee, Rohan Tikekar, Daniel Gallagher, James A. Lindsay, and The Listeria Market Basket Survey Multi-Institutional Team, and Sherri Dennis. 2017. Survey for *Listeria monocytogenes* in and on ready-to-eat foods from retail establishments in the United States (2010 through 2013): Assessing potential changes of pathogen prevalence and levels in a decade. *Journal of Food Protection* 80: 903–921.

Lund, Barbara M., and M.W. Peck. 1994. Heat resistance and recovery of spores of non-proteolytic *Clostridium botulinum* in relation to refrigerated, processed foods with an extended shelf-life. *Journal of Applied Bacteriology Symposium Supplement* 76: 115S–128S.

MacDonald, Emily, Berit Tafjord Heier, Karin Nygård, Torunn Stakheim Kofitsyo S. Cudjoe, Taran Skjerdal, Astrid Louise Wester, Bjørn-Arne Lindstedt, Trine-Lise Stavnes, and Line Vold. 2012. *Yersinia enterocolitica* outbreak associated with ready-to-eat salad mix, Norway, 2011. *Emerging Infectious Diseases* 18: 1496–1499.

MacDonald, E., M. Einöder-Moreno, K. Borgen, L. Thorstensen Brandal, L. Diab, Ø. Fossli, B. Guzman Herrador, A.A. Hassan, G.S. Johannessen, E.J. Johansen, R. Jørgensen Kimo, T. Lier, B.L. Paulsen, R. Popescu, C. Tokle Schytte, K. Sæbø Pettersen, L. Vold, Ø. Ørmen, A.L. Wester, M. Wiklund, and K. Nygård. 2016. National outbreak of *Yersinia enterocolitica* infections in military and civilian populations associated with consumption of mixed salad, Norway, 2014. Eurosurveillance 21:1–9. https://doi.org/10.2807/1560-7917.ES.2016.21.34.30321. Accessed 28 December 2017.

Marth, Elmer H. 1998. Extended shelf life refrigerated food: Microbiological quality and food safety. Food Technology. 52(2):57–62.

Morbidity Mortality Weekly Report (MMWR). 2003. *"Yersinia enterocolitica gastroenteritis among infants exposed to chitterlings—Chicago, Illinois, 2002."* Centers for Disease Control and Prevention. https://www.cdc.gov/mmwr/preview/mmwrhtml/mm5240a2.htm. Accessed 28 December 2017.

National Advisory Committee on Microbiological Criteria for Foods (NACMCF). 2005. Considerations for establishing safety-based consume-by date labels for refrigerated ready-to-eat foods. *Journal of Food Protection* 68: 1761–1775.

National Advisory Committee on Microbiological Criteria for Foods (NACMCF). 2010. Parameters for determining inoculated pack/challenge study protocols. *Journal of Food Protection* 73: 140–202.

New Zealand Food Safety Authority (NZFSA). 2005. *A Guide to Calculating the Shelf Life of Foods.* Wellington, NZ: New Zealand Food Safety Authority. http://blpd.dss.go.th/micro/A%20 Guide%20to%20Calculating%20the%20Shelf%20Life%20of%20Foods%20-%20New%20 Zealand.pdf. Accessed 28 December 2017.

Notermans, S., J. Dufrenne, P. Teunis, R. Beumer, M. te Giffel, and P. Peeters Weem. 1997. A risk assessment study of *Bacillus cereus* present in pasteurized milk. *Food Microbiology* 14: 143–152.

Peck, M.W. 2006. *Clostridium botulinum* and the safety of minimally heated, chilled foods: An emerging issue? *Journal of Applied Microbiology* 101: 556–570.

Pouillot, Regis, Daniel Gallagher, Jia Tang, Karin Hoelzer, Janell Krause, and Sherri B. Dennis. 2015. *Listeria monocytogenes* in retail delicatessens: An interagency risk assessment-model and baseline results. *Journal of Food Protection* 78: 134–145.

Pradhan, Abani K., R. Enata Ivanek, Yrjö T. Gröhn, Robert Bukowski, and Martin Wiedmann. 2011. Comparison of public health impact of *Listeria monocytogenes* product-to-product and environment-to-product contamination of deli meats at retail. *Journal of Food Protection* 74: 1860–1868.

Robins-Browne, Roy M. 2001. Yersinia enterocolitica. In *Food Microbiology Fundamentals and Frontiers*, ed. Michael P. Doyle, Larry R. Beuchat, and Thomas Montville, 2nd ed., 215–245. Washington, DC: ASM Press.

Scott, V.N., K.M.J. Swanson, T.A. Freier, W.P. Pruett, W.H. Sveum, P.A. Hall, L.A. Smoot, and D.G. Brown. 2005. Guidelines for conducting *Listeria monocytogenes* challenge testing of foods. *Food Protection Trends* 25: 818–825.

Szabo, Elizabeth, and Angela M. Gibson. 2003. Clostridium botulinum. In *Foodborne Microorganisms of Public Health Significance*, ed. Alisa D. Hocking, 6th ed., 505–542. Marrickville, NSW: Southwood Press.

Tacket, C.O., J. Ballard, N. Harris, J. Allard, C. Nolan, T. Quan, and M.L. Cohen. 1985. An outbreak of *Yersinia enterocolitica* infections caused by contamination tofu (soybean curd). *American Journal of Epidemiology* 121: 705–711.

Thompson, J. Stephen, and Michael J. Gravel. 1986. Family outbreak of gastroenteritis due to *Yersinia enterocolitica* serotype O:3 from well water. *Canadian Journal of Microbiology* 32: 700–701.

Todd, Ewen C.D., Judy D. Greig, Charles A. Bartleson, and Barry S. Michaels. 2007. Outbreaks where food workers have been implicated in the spread of foodborne disease. Part 3. Factors contributing to outbreaks and description of outbreak categories. *Journal of Food Protection* 9: 2199–2217.

Tompkin, R.B. 2002. Control of *Listeria monocytogenes* in the food processing environment. *Journal of Food Protection* 65: 709–725.

United States Department of Agriculture (USDA FSIS). 2011. *Yersiniosis and Chitterlings: Tips to Protect You and Those You Care for from Foodborne Illness.* Washington, DC: USDA FSIS. https://www.fsis.usda.gov/shared/PDF/Yersiniosis_and_Chitterlings.pdf. Accesses 28 December 2017.

United States Department of Agriculture, Food Safety and Inspection Service (USDA FSIS). 2010. *"FSIS Comparative Risk Assessment for Listeria monocytogenes In Ready-to-eat Meat and Poultry Deli Meats."* https://www.fsis.usda.gov/shared/PDF/Comparative_RA_Lm_Report_May2010.pdf. Accessed 28 December 2017.

United States Food and Drug Administration, Center for Applied Nutrition (FDA CFSAN). 2003. *"Quantitative assessment of relative risk to public health from foodborne Listeria monocytogenes among selected categories of ready-to-eat foods."* https://www.fda.gov/downloads/Food/FoodScienceResearch/UCM197330.pdf. Accessed 28 December 2017.

van Netten, P., A. van de Moosdijk, P. van Hoensel, D.A.A. Mossel, and I. Perales. 1990. Psychrotrophic strains of *Bacillus cereus* producing enterotoxin. *Journal of Applied Bacteriology* 69: 73–79.

Walker, S.J., P. Archer, and J.G. Banks. 1990. Growth of *Listeria monocytogenes* at refrigeration temperatures. *Journal of Applied Bacteriology* 68: 157–162.

Chapter 3
Microbial Growth and Spoilage

Peter J. Taormina

3.1 Introduction

Spoilage is characterized by any change in a food product that results in unacceptable sensory perception. In refrigerated foods, microbial growth often precedes or induces chemical and physical degradation. Hence, concern with microbial growth leading to spoilage receives higher priority than chemically or physically degradative processes. Growth of microorganisms in food systems is dependent on extrinsic and intrinsic factors. Extrinsic factors include temperature, atmosphere, exposure to light, and physical handling. Intrinsic factors include availability of nutrients, water activity (a_w), pH, presence of antimicrobial ingredients, competitive microflora, and endogenous enzymes. While chemical and physical degradation is certainly important and influenced by these factors, the rate of microbial growth on refrigerated foods leading to end of shelf life usually occurs more rapidly than autocatalytic chemical reactions. Consequently, change in microbial population is often used to assess and monitor shelf-life performance, with sensory evaluation always remaining the key aspect of shelf-life evaluation and ongoing testing. Microbial growth and metabolism in foods can impart biochemical changes (e.g., pH decline and production of acetoin) and possible formation of toxic metabolites, such as biogenic amines, or accumulation of odoriferous compounds, such as H_2S, byproducts of heterofermentative metabolism like CO_2, and exopolysaccharides (in't Veld 1996). Temperature has the greatest impact on the rate of deterioration of most perishable foods as increasing temperature increases the rates of reactions (Chandler and McMeekin 1989; Jacxsens et al. 2002; Ronsivalli and Charm 1975). A brief overview of quality deterioration reaction kinetics as affected by temperature was presented in Chap. 1, and growth of psychrotrophic bacterial pathogens was covered in Chap. 2. This chapter continues exploring

P. J. Taormina (✉)
Etna Consulting Group, Jacksonville, FL, USA
e-mail: peter@etnaconsulting.com

© Springer Nature Switzerland AG 2021
P. J. Taormina, M. D. Hardin (eds.), *Food Safety and Quality-Based Shelf Life of Perishable Foods*, Food Microbiology and Food Safety,
https://doi.org/10.1007/978-3-030-54375-4_3

temperature-dependent rates of spoilage as well as growth temperature ranges of various microorganisms, followed by a review of spoilage microflora in various food categories.

3.2 Role of Temperature in Shelf Life

Temperature has a profound impact upon the rate of chemical, physical, and microbial deterioration of foods. There are several mathematical models to estimate the deterioration rate of foods during storage as affected by temperature as well as other factors (Mizrahi 2004). The Arrhenius equation $k = A\exp\left(-\dfrac{Ea}{RT}\right)$ linearly describes many different types of reactions, including shelf life. The approach recommended for shelf-life testing is to assume that certain principles of chemical kinetics apply with respect to temperature acceleration, as described by the Arrhenius relationship, and utilize kinetic design to make accurate shelf-life predictions (Labuza 1984). Peleg et al. (2012) pointed out that the apparent linearity of the Arrhenius plot in many food systems is due to a mathematical property of the model's equation rather than to the existence of a temperature-independent "energy of activation" and proposed an exponential model that better describes temperature dependencies traditionally described by the Arrhenius equation without the assumption of a temperature-independent "energy of activation." Stannard et al. (1985) studied the growth of psychrotrophic spoilage bacteria in pure culture and determined that square root relationship ($\sqrt{r} = b(T - T_o)$) was better than the Arrhenius equation for the description of the microbial growth/temperature relationship at chill temperatures and was also applicable to mixtures of organisms.

The complex interactions of microbial metabolism and impact on food biochemistry as it relates to temperature have been long studied (Mossel and Ingram 1955). Foundational knowledge about microbial growth rates was advanced to the point of predicting microbial growth as affected by temperature and other factors (Baranyi and Roberts 1994; Wilson et al. 2002). Such predictions are useful especially when temperature history of raw materials and ingredients, work-in-process food, and finished food products varies. As such, the microbial profile of the raw materials and finished product can change throughout the production process and impact the degree of chemical and physical degradation and ultimately affect the shelf life. Indeed, temperature changes infuse a high degree of complexity in predicting shelf life. Such complexity lends itself to predictive modeling, but such models are based on mathematical equations developed at many static temperatures that must be validated with experimental data.

Research on microbial metabolism and physiochemical breakdown of food substrates was prompted by a goal to develop reliable time–temperature indicators for the purpose of shelf-life prediction (Riva et al. 2001; Vaikousi et al. 2008). While application of such predictive models in the day-to-day operations within the food industry is not widespread, opportunities do occasionally arise for industry collaboration with academia or government to utilize shelf-life prediction. For example, predictive models for shelf life may be utilized for new product lines or for major initiatives to extend shelf life of existing product lines.

3.3 Storage Temperatures and Microbial Growth

Often, the storage temperature for raw materials or finished products is based on the minimum growth temperatures of microorganisms that are likely to grow and spoil the food. The minimum growth temperatures of selected bacteria are displayed in Table 3.1. As can be expected, the minimum growth temperatures widely range based on bacterium and growing conditions, but notably *Pseudomonas*, *Brochothrix*, *Lactobacillus*, *Lactococcus*, and *Weissella* have been demonstrated to grow below 0 °C.

The minimum growth temperatures of selected fungi are displayed in Table 3.2. Molds with ability to grow below 0 °C included *Aureobasidium pullulans*, *Botrytis cinerea*, *Cladosporium* spp., *Fusarium sporotrichioides*, and *Penicillium* spp. Fewer yeasts were found to grow at negative temperatures, but *Rhodotorula glutinis* was found to grow on blanched peas at as low as −18 °C.

Bacterial pathogens tend to have higher minimum growth temperatures than either spoilage bacteria or fungi (Table 3.3). This is of great practical importance as growth of these pathogens will often lag that of the nonpathogenic, psychrotrophic spoilage flora of foods. Of the pathogens, psychrotrophic *Bacillus cereus*, nonproteolytic *Clostridium botulinum*, *Listeria monocytogenes*, and *Yersinia enterocolitica* are capable of growth at the lowest temperatures. These pathogens were addressed in detail in Chap. 2.

3.3.1 Freezing Temperatures

Microbes survive in frozen environments with little nutrient, like glacial and sea ice and permafrost, at temperatures well below the freezing point of water (Mackelprang et al. 2017). Microbial cells surviving in such harsh frozen environments are largely dormant but can repair macromolecular damage by means of DNA-repair enzymes and protein repair enzymes (Price and Sowers 2004). Microorganisms maintain viability in frozen foods for long periods of time. When cells were suspended in 10% glycerol and stored at −53 °C for 16 months, there were some species and genera differences in survivability (Yamasato et al. 1973). Strains of coryneform bacteria, genera of the family *Enterobacteriaceae*, and the genus *Pseudomonas* showed relatively higher survivability, whereas organisms such as *Pseudomonas putrefaciens* were sensitive. Lactic acid bacteria became sublethally injured and required special resuscitation procedures following frozen storage, but all acetic acid bacteria sur-

Table 3.1 Minimum growth temperature (°C) of selected bacteria

Organism	Minimum temperature	Substrate	pH	Atmosphere	Temperature condition	Refs.
Acetobacter	5	–	–	–	–	Kraft (1992)
Acinetobacter	4	–	–	–	–	Kraft (1992)
Aeromonas	5	–	–	–	–	Kraft (1992)
Aeromonas hydrophila	Var., ≤0–10	Tryptic Soy Broth (TSB)	–	Aerobic	Isothermal	Rouf and Rigney (1971)
Aeromonas salmonicida	Var., ≤5–10	TSB	–	Aerobic	Isothermal	Rouf and Rigney (1971)
Aeromonas shigelloides	Var., ≤0–10	TSB	–	Aerobic	Isothermal	Rouf and Rigney (1971)
Arthrobacter	5	–	–	–	–	Kraft (1992)
Brevibacterium	5	–	–	–	–	Kraft (1992)
Brochothrix thermosphacta	–3.36	Modified Elliker broth	Opt (ca 7–7.4)	–	Isothermal	Leroi et al. (2012)
Chromobacterium	2	–	–	–	–	Kraft (1992)
Chromobacterium fluviatile	4	–	–	–	–	Kraft (1992)
Chromobacterium lividum	2	–	–	–	–	Kraft (1992)
Clostridium putrefaciens	0	–	–	–	–	Kraft (1992)
Cytophaga xantha	<0	–	–	–	–	Kraft (1992)
Escherichia coli	5–10	–	–	–	–	Kraft (1992)
Flavobacterium	5	–	–	–	–	Kraft (1992)
Gluconobacter oxydans	7	–	–	–	–	Kraft (1992)
Kurthia	5	–	–	–	–	Kraft (1992)
Lactobacillus	2	–	–	–	–	Kraft (1992)
Lactobacillus[a]	<–1	DeMan-Rogosa-Sharpe (MRS) agar	–	–	Isothermal	Korkeala et al. (1990)

Organism	Minimum temperature	Substrate	pH	Atmosphere	Temperature condition	Refs.
Lactobacillus curvatus	−3.27 ± 1.21	MRS broth	5.5–8.5[b]	Aerobic	–	Wijtzes et al. (1995)
Lactobacillus pentosus	>4	–	–	–	–	Lund et al. (2000)
Lactobacillus plantarum	>4	–	–	–	–	Lund et al. (2000)
Lactobacillus plantarum	3.29	MRS broth	–	–	–	Zwietering et al. (1994)
Lactococcus piscium	−4.8	Modified Elliker broth	Opt (ca 7–7.4)	–	Isothermal	Leroi et al. (2012)
Leuconostoc	5	–	–	–	–	Kraft (1992)
Leuconostoc mesenteroides	4, 3[c]	MRS agar	–	–	Isothermal	Korkeala et al. (1990)
Microbacterium	≤10	–	–	–	–	Kraft (1992)
Moraxella	2	–	–	–	–	Kraft (1992)
Pediococcus	>4	–	–	–	–	Lund et al. (2000)
Propionibacterium	2–3	–	–	–	–	Kraft (1992)
Pseudomonas	−12.8, −11.8[d]	Fish (gilthead seabream)	–	Aerobic	Nonisothermal	Koutsoumanis (2001)
Pseudomonas	4	–	–	–	–	Kraft (1992)
Pseudomonas	−0.08	Beef muscle	–	–	–	Zhang et al. (2011)
Pseudomonas fluorescens	0–4	–	–	–	–	Kraft (1992)
Pseudomonas putida	≤0	Histidine medium	–	–	Isothermal	Hug and Hunter (1974)
Serratia	4–5	–	–	–	–	Kraft (1992)
Staphylococcus	5–10	–	–	–	–	Kraft (1992)
Streptococcus faecalis	5–10	–	–	–	–	Kraft (1992)
Weissella cibaria	9.5 ± 1.5	Weissella medium	6.5	–	–	Ricciardi et al. (2009)
Weissella viridescens	−1.33 ± 1.26[e]	MRS broth	–	–	Isothermal	Longhi et al. (2016)
Weissella viridescens	0.12 ± 0.71[f]	MRS broth	–	–	Nonisothermal	Longhi et al. (2016)

(continued)

Table 3.1 (continued)

Organism	Minimum temperature	Substrate	pH	Atmosphere	Temperature condition	Refs.
Weissella viridescens	-1.57 ± 1.05^g	MRS broth	–	–	Nonisothermal	Longhi et al. (2016)
Xanthomonas	>5	–	–	–	–	Kraft (1992)

[a]Ropy slime-producing, homofermentative lactobacilli

[b]T_{min} was independent of pH

[c]Continuous growth decreased at ca. 4 °C; microcolonies not observed microscopically below 3 °C

[d]T_{min} parameters for Belehradek models for t_{Lag} and μ_{max}, respectively

[e]Two-step modeling approach

[f]Increasing temperature optimal experimental design approach

[g]Decreasing temperature optimal experimental design approach

Table 3.2 Minimum growth temperature (°C) of selected fungi

Microorganism	Minimum temperature	Substrate	pH	Water activity	Refs.
Molds					
Alternaria alternata	Var., −5–6.5	–	–	–	Pitt and Hocking (1997)
Aspergillus candidus	3–4	–	–	–	Pitt and Hocking (1997)
Aspergillus clavatus	5–6	–	–	–	Pitt and Hocking (1997)
Aspergillus flavipes	6–7	–	–	–	Pitt and Hocking (1997)
Aspergillus flavus	Var., 10–12	–	–	–	Pitt and Hocking (1997)
Aspergillus fumigatus	Ca. 12	–	–	–	Pitt and Hocking (1997)
Aspergillus niger	10.13	Malt extract agar	4.2	0.997	Gougouli and Koutsoumanis (2010)
Aspergillus niger	6–8	–	–	–	Pitt and Hocking (1997)
Aspergillus ochraceus	8	–	–	–	Pitt and Hocking (1997)
Aspergillus versicolor	9	–	–	0.97	Pitt and Hocking (1997)
Aureobasidium pullulans	−5	–	–	–	Pitt and Hocking (1997)
Aureobasidium pullulans var. pullulans	4	–	–	–	Samson et al. (2010)
Aureobasidium pullulans var. melanogenum	10	–	–	–	Samson et al. (2010)
Absidia corymbifera	14	–	–	–	Pitt and Hocking (1997)
Botrytis cinerea	Var., −2–12	–	–	–	Pitt and Hocking (1997)
Byssochlamys fulva	6.26	Solidified apple juice	–	–	Longhi et al. (2014)
Byssochlamys fulva	9.1	Malt extract agar	–	–	Panagou et al. (2010)
Byssochlamys nivea	10.5	Malt extract agar	–	–	Panagou et al. (2010)
Chaetomium globosum	4–10	–	–	–	Pitt and Hocking (1997)
Cladosporium cladosporioides	−5	–	–	–	Pitt and Hocking (1997)
Cladosporium herbarum	−10	–	–	–	Pitt and Hocking (1997)

(continued)

Table 3.2 (continued)

Microorganism	Minimum temperature	Substrate	pH	Water activity	Refs.
Cladosporium herbarum	Ca. –5	Meat	–	–	Pitt and Hocking (1997)
Endomyces fibuliger	Ca. 5	–	–	–	Pitt and Hocking (1997)
Emericella nidulans	6–8	–	–	–	Pitt and Hocking (1997)
Epicoccum nigrum	Var., <5	–	–	–	Pitt and Hocking (1997)
Eupenicillium cinnamopurpureum	4–6	–	–	–	Pitt and Hocking (1997)
Eurotium repens	4–5	–	–	–	Pitt and Hocking (1997)
Eurotium rubrum	Ca. 5	–	–	–	Pitt and Hocking (1997)
Fusarium culmorum	0	–	–	–	Pitt and Hocking (1997)
Fusarium moniliforme	2.5–5	–	–	–	Pitt and Hocking (1997)
Fusarium oxysporum	> 5	–	–	–	Pitt and Hocking (1997)
Fusarium poae	Ca. 2.5	–	–	–	Pitt and Hocking (1997)
Fusarium sporotrichioides	–2	–	–	–	Pitt and Hocking (1997)
Lasiodiplodia theobromae	15	–	–	–	Pitt and Hocking (1997)
Monascus ruber	14·61 ± 1·92	Malt extract agar	–	–	Panagou et al. (2003)
Monascus ruber	14.91 ± 1.85	Malt extract agar	–	–	Panagou et al. (2003)
Monascus ruber	15–18	–	–	–	Pitt and Hocking (1997)
Mucor hiemalis	<0	–	–	–	Pitt and Hocking (1997)
Mucor piriformis	Ca. 0	–	–	–	Pitt and Hocking (1997)
Mucor racemosus	Ca. –4	–	–	–	Pitt and Hocking (1997)
Neosartorya fischeri	11–13	–	–	–	Samson et al. (2010)
Paecilomyces variotii	Ca. 5	–	–	–	Pitt and Hocking (1997)
Penicillium aurantiogriseum	Ca. –2	–	–	–	Pitt and Hocking (1997)

Table 3.2 (continued)

Microorganism	Minimum temperature	Substrate	pH	Water activity	Refs.
Penicillium brevicompactum	−2	−	−	−	Pitt and Hocking (1997)
Penicillium chrysogenum	4	−	−	−	Pitt and Hocking (1997)
Penicillium citrinum	Ca. 5	−	−	−	Pitt and Hocking (1997)
Penicillium crustosum	Ca. −2	−	−	−	Pitt and Hocking (1997)
Penicillium expansum	−5.74	Malt extract agar	4.2	0.997	Gougouli and Koutsoumanis (2010)
Penicillium expansum	−6, −3, −2[a]	−	−	−	Pitt and Hocking (1997)
Penicillium funiculosum	Ca. 8	−	−	−	Pitt and Hocking (1997)
Penicillium glabrum	≤0	−	−	−	Pitt and Hocking (1997)
Penicillium hirsutum	−5	−	−	−	Pitt and Hocking (1997)
Penicillium islandicum	10	−	−	−	Pitt and Hocking (1997)
Penicillium italicum	−3, 0	−	−	−	Pitt and Hocking (1997)
Penicillium oxalicum	8	−	−	−	Pitt and Hocking (1997)
Penicillium purpurogenum	12	−	−	−	Pitt and Hocking (1997)
Penicillium variabile	12	−	−	−	Pitt and Hocking (1997)
Penicillium viridicatum	<−2	−	−	−	Pitt and Hocking (1997)
Rhizomucor miehei	21	−	−	−	Samson et al. (2010)
Rhizomucor pusillus	20	−	−	−	Pitt and Hocking (1997)
Rhizopus oligosporus	12	−	−	−	Samson et al. (2010)
Rhizopus oryzae	7	−	−	−	Pitt and Hocking (1997)
Rhizopus oryzae	5–7	−	−	−	Samson et al. (2010)
Rhizopus stolonifer	4.5–5	−	−	−	Pitt and Hocking (1997)
Rhizopus stolonifer	Ca. 5	−	−	−	Samson et al. (2010)

(continued)

Table 3.2 (continued)

Microorganism	Minimum temperature	Substrate	pH	Water activity	Refs.
Scopulariopsis candida	5	–	–	–	Samson et al. (2010)
Scopulariopsis fusca	5	–	–	–	Samson et al. (2010)
Thamnidium elegans	≤1	–	–	–	Pitt and Hocking (1997)
Wallemia sebi	>5	–	–	–	Samson et al. (2010)
Yeasts					
Candida krusei	Ca. 8	–	–	–	Pitt and Hocking (1997)
Candida parapsilosis	Ca. 8	–	–	–	Pitt and Hocking (1997)
Debaryomyces hansenii	Ca. 2	YM broth	–	–	Pitt and Hocking (1997)
Kloeckera apiculata	Ca. 8	–	–	–	Pitt and Hocking (1997)
Pichia membranaefaciens	Ca. 5	–	–	–	Pitt and Hocking (1997)
Rhodotorula mucilaginosa	0.5–5	–	–	–	Pitt and Hocking (1997)
Rhodotorula glutinis	<−18	Blanched peas			Pitt and Hocking (1997)
Saccharomyces cerevisiae					Pitt and Hocking (1997)
10% Glucose	4	–	'	–	–
50% Glucose	13	–		–	–
Trichoderma harzianum	Ca. 5	–	–	–	Pitt and Hocking (1997)
Trichothecium roseum	Ca. 5, 15[b]	–	–	–	Pitt and Hocking (1997)
Zygosaccharomyces bailii					Pitt and Hocking (1997)
10% Glucose	6.5	–	–	–	
30% Glucose	6.5	–	–	–	
60% Glucose	13	–	–	–	
Zygosaccharomyces rouxii					Pitt and Hocking (1997)
10% Glucose	< 4	–	–	–	
60% Glucose	7	–	–	–	

[a]Minimum temperature reported as −6 °C (Brooks and Hansford 1923), −3 °C (Panasenko 1967), and − 2 °C (Mislivec and Tuite 1970b)
[b]Minimum temperature reported as ca. 5 (Pitt and Hocking 1997) and 15 (Domsch et al. 1980)

Table 3.3 Minimum growth temperature (°C) of selected pathogens

Microorganism	Minimum temperature	Substrate	Refs.
Bacillus cereus, psychrotrophic	4	–	ICMSF (1996)
Bacillus cereus, mesophilic	15	–	ICMSF (1996)
Campylobacter jejuni	32[a]	Laboratory media	ICMSF (1996)
Clostridium botulinum, nonproteolytic	3.3	–	ICMSF (1996)
Clostridium botulinum, proteolytic	10	–	ICMSF (1996)
Clostridium perfringens	12	–	ICMSF (1996)
E. coli O157:H7	8	–	ICMSF (1996)
E. coli, pathogenic	7	–	ICMSF (1996)
Listeria monocytogenes	Ca. 0	High nutrient content medium	ICMSF (1996)
Listeria monocytogenes	−0.4	–	ICMSF (1996)
Salmonella[b,c]	5.2	–	ICMSF (1996)
Salmonella	5–10	–	Kraft (1992)
Shigella sonnei	6.1	–	ICMSF (1996)
Shigella flexneri	7.9	–	ICMSF (1996)
Staphylococcus aureus	7	–	ICMSF (1996)
Staphylococcus aureus, toxin production	10	–	ICMSF (1996)
Staphylococcus aureus	5–10	–	Kraft (1992)
Streptococcus pyogenes	10–15	–	ICMSF (1996)
Vibrio	4	–	Kraft (1992)
Vibrio cholerae	10	–	ICMSF (1996)
Vibrio parahaemolyticus	5	–	ICMSF (1996)
Vibrio vulnificus	8	–	ICMSF (1996)

(continued)

Table 3.3 (continued)

Microorganism	Minimum temperature	Substrate	Refs.
Yersinia enterocolitica	−1.3	Laboratory media,optimal nutrient conditions	ICMSF (1996)
Yersinia enterocolitica	0–4	–	Kraft (1992)

[a]Microaerobic atmosphere
[b]Most serotypes fail to grow at <7 °C. Growth rate substantially reduced at <15 °C
[c]Ability of *Salmonella* to grow at temperatures <5 °C has been reported (D'Aoust 1991)

vived well in 10% glycerol and in 10% honey at −28 °C for 4.5 years. Interestingly, the authors determined that 10% honey was a better cryoprotectant than 10% glycerol. Frozen foods tend to be stored at −20 °C, which is below the minimum for growth of the organisms reviewed in Tables 3.1, 3.2, and 3.3. However, temperature settings on freezers utilized in the cold chain of food distribution, storage, and display may cause temperatures to rise closer to 0 °C, thereby permitting slow metabolic activity and some growth of certain microorganisms.

3.3.2 Refrigeration Temperatures

What is considered refrigerated storage temperatures for food also varies. Typically, storage temperatures used for raw meats, poultry, and seafood are just below freezing point of the muscle (ca. −7.2 °C). Food processing plants may, however, hold raw proteins at higher temperatures such as 4–10 °C for short periods of time when in-process. Refrigerated foods and ingredients are likely to be stored and distributed at temperatures within a range of 1.67–4.4 °C. At retail and foodservice settings, the maximum allowable temperature of storage is typically 5 °C. As can be seen from Tables 3.1, 3.2, and 3.3, many types of bacteria and fungi will readily grow at these temperatures. Refrigerated foods are sometimes subjected to moderately abusive temperatures upward of 10–12 °C during distribution, storage, and display. Exposure to such temperatures may be brief causing little to no change to the food temperature or can be long subjecting the food to more rapid microbial growth rates. Temperature cycling occurs when food is moved from one area to another for the purpose of distribution and/or display. The accuracy of shelf-life estimation is diminished the more, these temperature cycling scenarios are in play.

Psychrotrophic microorganisms tend to dominate the microbiota of refrigerated foods. Bacteria such as *Brochothrix thermosphacta*, *Shewanella* spp., *Aeromonas*, and/or *Enterobacteriaceae* can grow under refrigeration temperatures (Gram et al. 2002; Holley et al. 2004). *Pseudomonas* spp. and a few other gram-negative psychrotrophic organisms will dominate proteinaceous foods stored aerobically at chill temperatures (Gram et al. 2002), including meat, poultry, milk, and fish. Gram-negative bacteria like *Pseudomonas*, *Hafnia*, *Shewanella*, *Photobacterium*, and

Aeromonas grow in aerobically stored refrigerated products and degrade proteins, leading to formation of off odors and flavors. Gram-positive bacteria like homofermentative lactic acid bacteria, heterofermentative lactics, *Carnobacterium*, and *Brochothrix* reach 10^6–10^8 CFU/g during refrigerated storage. While strict aerobes like *Brochothrix thermosphacta* can be controlled by packaging film with low-oxygen permeability (Holley 1999), facultative anaerobes must be limited by sanitation, temperature control, and other factors intrinsic to the food, such as pH. Yeast and mold also readily grow on perishable foods, whether refrigerated or shelf stable. *Aspergillus*, *Penicillium*, *Mucor*, and *Rhizopus* are molds commonly causing food spoilage under refrigeration. Some of the more common yeasts genera that spoil refrigerated foods are Candida, Dekkera, and *Rhodotorula*.

3.4   Shelf-Life Determination

Microbiological testing has historically been used in the food industry to assess shelf life of raw materials and finished products. End of shelf life is often attributed to the food attaining 10^6–10^7 CFU/g. To some degree, this is due to habit and convenience. It can be easier to train technicians to perform simple microbiological assays than to train technicians to perform objective sensory evaluation of shelf-life samples. The methods for microbiological data collection and reporting are more streamlined compared to sensory analysis of products for shelf-life testing. In food processing, the shelf-life measurement methods often rely on the total plate count (TPC), which is essentially an aerobic mesophilic plate count. Other similar assays are aerobic plate count (APC) and standard plate count (SPC). Typically, agar media or film media are used for these purposes, and ISO methods for APC or TPC call for incubation of plates at 35 ± 1 °C and counting within 48 ± 3 h (ISO 4833:2003 and ISO 15214:1998). However, these incubation conditions do not permit maximum colony development for many psychrotrophic bacteria that are the most abundant mid to late shelf life. Therefore, the term, "Total," is a misnomer, and over reliance on this standard technique can result in incomplete data concerning the true microbial profile of food samples. A considerable underestimation (+0.5–3.2 log CFU/g) on the contamination levels of psychrotrophic lactic acid bacteria (LAB) was observed for retail, packaged food products stored at chilling temperature when the mesophilic enumeration technique was implemented as reference shelf-life parameter (Pothakos et al. 2012; Pothakos et al. 2014). Most of the isolates belonged to the genera *Leuconostoc*, *Lactococcus*, and *Lactobacillus* and were unable to grow at 30 °C. However, several strains of *Leuconostoc* can dominate the population of various packaged, refrigerated foods near the end of shelf life (Pothakos et al. 2014). In an acetic acid herring preserve product, the microbial diversity of the dominant psychrotrophic LAB recovered after incubation of plates at 22 °C for 5 days was determined using a polyphasic taxonomic approach (Lyhs et al. 2004). A total of 212 LAB isolates were identified using a combination of rep-PCR fingerprinting, amplified fragment length polymorphism

(AFLP) analysis, and pheS gene sequencing. *Leuconostoc gasicomitatum*, *Leuconostoc gelidum*, *Leuconostoc* spp., *Lactococcus piscium,* and *Lactobacillus algidus* proved to be the most competent and predominant species that may go undetected by the widely applied mesophilic enumeration protocols.

Microbiological data may be easy to gather and tabulate, but interpretation can be difficult, and false assumptions are often made about the data. For example, a simple threshold limit of population of APC such as 10^6 CFU/g may not correlate with sensory unacceptance. As for ongoing quality-based shelf-life assessments, the APC and/or lactic plate count data on samples will often exceed 10^6 or even 10^7, yet samples will look, smell, and taste acceptable. To avoid this confusion, some companies have decided that shelf-life end-point determinations should be based solely on visual/odor/flavor evaluations, knowing that from time to time, product could exceed 10^6 or even 10^7 of at around mid to late shelf life (e.g., 70–100 days). If reasonable empirical data can be gathered to demonstrate that organoleptic evaluations will capture microbial spoilage and "undesirable" microorganisms, then it would lend credence to essentially ignoring (not collecting data on) APC and lactic even on ready-to-eat (RTE) products that could support growth of psychrotrophic pathogens.

Beyond the standard plate counting techniques most used in industrial food microbiology laboratories, university and government laboratories often utilize techniques like impedance, fluorometry, and RT-PCR (Guy et al. 2006; Martínez et al. 2011), which are capable of rapid quantification in some food systems. Visible and short-wavelength near-infrared (SW-NIR) diffuse reflectance spectroscopy (600–110 nm) to quantify the microbial loads in chicken breast muscle and to develop a rapid methodology for monitoring the onset of spoilage and in conjunction with principal component analysis could detect APC increases slightly above 1-log (Lin et al. 2004). These techniques can be utilized for more rapid quantification of microorganisms for the purposes of measuring quality throughout shelf life but are typically cost prohibitive, and so the plate count or version of the plate count often prevails.

3.5   Accelerated Shelf-Life Testing for Refrigerated Foods

Normal shelf-life testing must span the stated or potential usable time that a food product is saleable and consumable, and ideally, the testing timeframe should exceed that target shelf life so that the point at which the product loses acceptability can be determined. However, the time from the genesis of an idea or stated goal to launch a food product into the marketplace to the time that it is commercialized will often exceed the desired shelf life of the product. When the ideation process has led to a food product development project, there should be a general idea of the desired number of days or weeks of shelf life for such product, but this number could be a rough estimate. For truly innovative new product development, the shelf-life target may be truly unknown. In either case, since products must have a stated shelf life prior to commercialization, the answer as to what that shelf life might be is often sought out as quickly as possible. Ergo, accelerated shelf-life testing has been utilized as means to predict the shelf life of perishable commodities such as refrigerated dairy or meat products (Corradini and Peleg 2007). Accelerated shelf-life testing is based on measurements of the rate constant (k) at several different temperatures, followed by extrapolation of the straight-line plot to lower temperatures for predicted shelf life (Man 2002).

Although accelerated shelf-life testing is often used by food laboratories to provide customers with the much-desired early result for shelf-life testing, the assumed relationship between temperature and reaction rate as it pertains to shelf life should be validated per product type. There is a formula based on the Arrhenius equation, which assumes a straight-line relationship between temperature and rate of degradation of product quality (i.e., end of shelf life). This is an equation based on "the rule of 10s" which was proposed by Ted Labuza in the 1980s, and has caught on for use by contemporary food microbiology testing labs:

$$Q10 = \frac{\text{Rate at temeprature} \left(T+10\right)^\circ \text{C}}{\text{Rate at temperature } T^\circ \text{C}} = \frac{\text{Shelf life at } T^\circ \text{C}}{\text{Shelf life at} \left(T+10\right)^\circ \text{C}}$$

This equation works nicely for shelf life of foods that is determinative based on some single chemical reaction like lipid oxidation or bread staling, but when applied to complex microbial communities, it loses its applicability. This is because the preferred growth temperatures of microorganisms are varied across genera. Indeed, elevated storage temperature of food selects for mesophilic microorganisms, putting psychrotrophs at a distinct disadvantage in competing for nutrients and physical space within or on a food substrate. Therefore, incubating product samples at room temperature (ca. 22 °C) and correlating the time to spoilage with an actual refrigerated product shelf life is dubious without rigorous data to establish the correlation between shelf life at both elevated and refrigerated temperatures. A multivariate accelerated shelf-life test for fresh-cut lettuce was developed using principal component analysis and leading to shelf life of 12.4, 10.4, and 3.7 days at 0, 5, and 15 °C (Derossi et al. 2016). However, only sensory, physical, and chemical attributes were assessed, not microbial. A similar study design and outcome was reported for pineapple (Amodio et al. 2015). A study of accelerated shelf-life testing of ice cream utilized total aerobic counts as well as sensory and pH in the Arrhenius equation to estimate shelf life (Park et al. 2018), but the study did not explain growth of aerobic counts at very low temperatures such as −6 °C, which is below the minimum growth temperature for most microorganisms in foods (Tables 3.1, 3.2, and 3.3).

3.6 Chemical Indicators of Microbial Growth

Chemical changes to food during shelf life include oxidation of lipids and pigments, protein denaturation, and fermentation of carbohydrates. A shift in titratable acidity or pH can be indicative of growth of acid-producing microorganisms such as

Lactobacillus, *Acetobacter*, or *Gluconobacter*. Chemical indicators of spoilage such as D-lactate, in foods that do not contain that molecule, or pH reduction can be correlated with microbial growth and impact on shelf life. In vacuum-packaged pork, D(−)lactic acid and acetoin/diacetyl concentrations increased progressively although microbial counts stabilized from the 20th day of refrigerated storage (Pablo et al. 1989). Several vacuum-packed meat products from the market had D(−)lactic acid concentrations that were indicative of the sensory and microbial end of shelf-life target. In raw pork stored under high-oxygen modified atmosphere at +4 °C levels of acetoin, diacetyl and 3-methyl-1-butanol correlated with the sensory scores and bacterial concentrations (Nieminen et al. 2016). Acetoin production can be associated with growth of various types of bacteria including lactic acid bacteria, *Listeria monocytogenes*, and *Bacillus* spp. (El-Gendy et al. 1983; Xiao and Lu 2014; Zeppa et al. 2001).

Biogenic amines histamine, putrescine, cadaverine, spermine, spermidine, and tyramine are toxic metabolites primarily produced by metabolically active *Enterobacteriaceae* and enterococci. Lactic acid bacteria do not typically produce significant levels of biogenic amines except during fermentation of meat and dairy products. Biogenic amines are mainly formed by decarboxylation of amino acids. *Enterobacteriaceae*, heterofermentative lactic acid bacteria, *Enterococcus faecalis*, and *Lactobacillus buchneri* are among the bacteria capable of producing more than 500 ppm of biogenic amines during meat or cheese fermentation (Janssen et al. 1996). With respect to shelf-life, biogenic amine production would be a concern for fresh meat, poultry, and seafood, and less so if temperature control is maintained.

Trimethylamine oxide reductase enzymes are utilized by bacteria to reduce trimethylamine oxide (TMAO) to the trimethylamine (TMA), an odorous substance. N-acyl homoserine lactones (AHL) are produced by a wide variety of bacteria growing in several different food substrates apparently as a means for bacterial quorum sensing (Gram et al. 2002). Quantities of AHL may correlate with the onset of sensory decomposition of foods and could therefore be a potential spoilage indicator. AHL can be extracted from food by homogenizing with ethyl acetate and can be visualized by separation of extracted AHLs on thin-layer chromatographic plates and subsequent development by AHL-monitor strains, for example, *Agrobacterium tumefaciens* pZLR4 (Gram et al. 2002). Levels of volatile organic compounds such as aldehydes, ketones, and alcohols can indicate microbiological spoilage of beef (Saraiva et al. 2015), pork (Nieminen et al. 2016), poultry (Tománková et al. 2012), seafood (Mikš-Krajnik et al. 2016), dairy products (Condurso et al. 2008), and tropical fruits (Taiti et al. 2015) and iceberg lettuce (Ioannidis et al. 2018) to name a few.

Thiobarbituric reactive substances (TBARS) are commonly used in meat and poultry to determine lipid oxidation, mostly not related to microbial growth. However, a combination of microbiological and chemical spoilage markers can lead to better shelf-life estimation. In cooked blood sausage, quantitative descriptive analysis, *Pseudomonas* spp., lactic acid bacteria and *Enterobacteriaceae* counts, TBARS, total basic volatile nitrogen, and lactic acid showed a better predictive ability of consumer acceptability than any single spoilage indicators (Pereira et al. 2019).

3.7   Shelf Life of Specific Food Types

3.7.1   *Fresh Produce*

Fresh produce has a wide variety of intrinsic properties that have a bearing on microbial growth rate. Many vegetables tend to have a near neutral pH, while many fruits tend to be acidic. Most but not all produce has substantial moisture content and carbohydrate. Most fresh produce contains only trace amounts of protein and can vary in lipid content from significant (avocados) to trace (lettuce). Various raw agricultural produce items possess a natural barrier in the form of a rind or "skin" that protects the fleshy trace tissues from physical damage, oxidation of pigments and nutrients, and microbial infiltration. Temperature and microflora are the key factors affecting shelf life of produce (Brackett 1987). Microorganisms that typically grow on and eventually spoil whole fresh produce include *Enterobacteriaceae*, pseudomonads, lactic acid bacteria, yeasts, and molds.

Once produce is cut and fleshy tissue is exposed, growth rates increase, and shelf life shortens. A survey of 120 Australian retail samples of trimmed, cut lettuce usually with several days of remaining shelf life revealed most samples to have APC in the range of 10^5 and 10^7 CFU/g (Szabo et al. 2000). Bacterial pathogens *Listeria monocytogenes*, *Aeromonas* spp., and *Yersinia enterocolitica* were isolated from 3, 66, and 71 samples, respectively. These pathogens grow readily at refrigeration temperatures (Tables 3.1 and 3.3). In a shelf-life study of cut fruit pieces stored at 4 °C, it was found that the signs of spoilage differed among fruit types (O'Connor-Shaw et al. 1994). Brown discoloration was observed only in pineapple, while bitter flavor developed only in kiwifruit. Microbial growth did not appear to contribute to spoilage in diced kiwifruit, papaya, and pineapple, but may have affected honeydew melon and cantaloupe. No group of microorganisms (lactobacilli, *Enterobacteriaceae*, yeasts, and molds) dominated the microflora of fresh minimally processed fruit regardless of species of fruit (O'Connor-Shaw et al. 1994). Kiwifruit, papaya, and pineapple pieces became softer during storage. Undesirable textural changes in cut fruit are caused by enzymes such as β-galactosidase and exo-polygalacturonase, which solubilize pectin in cell walls (King and Bolin 1989).

Shelf life of fresh processed vegetables is increased with the use of modified atmosphere packaging (MAP). Due to senescence, the modified atmospheres generated inside packages initially can change over the course of shelf life and can contain moderate-to-high levels of CO_2. While CO_2 restricts the growth of aerobic microorganisms, it can lead to safety concerns from anaerobic sporeforming bacteria such as *Clostridium botulinum*. The lack of growth of aerobic microorganisms due to MAP coupled with potential *C. botulinum* growth could mean that the organoleptic evidence of microbial growth could be lacking despite a pathogen prevalence (Farber 1991). MAP is widely used to extend the refrigerated shelf life of fresh cut produce. Modified atmosphere stored mixed lettuce and cucumber slices had degraded sensory properties that preceded microbial proliferation and achieved 7 days at 4 °C, but microbial proliferation was the limiting factor for cut bell peppers (Jacxsens et al. 2002). The product pH, starting microbial load, and storage temperature impacted growth of *L. monocytogenes* and *Aeromonas caviae* on lettuce and cucumber.

3.7.2   Pasteurized Milk and Dairy Products

Milk is an ideal growth medium for microorganisms with ample protein, lipids, and neutral pH. Consequently, shelf life of milk can be very short if microorganisms are present and temperatures approach 10 °C. In Europe and North America for instance, pasteurized milk must be heated at high temperature for a short time (at least 71.7 °C for 15 s, or any equivalent combination) and must show a negative reaction to the phosphatase test and a positive reaction to the peroxidase test. Immediately after pasteurization, milk must be cooled to 6 °C or below. The shelf life of pasteurized milk is dependent on raw milk quality, pasteurization conditions, contamination from the food contact surface, contamination from environment, distribution temperature, and effect of light (Rysstad and Kolstad 2006). Psychrotrophic microorganisms are the primary cause of pasteurized milk spoilage and the determining factor in shelf life (Chandler and McMeekin 1989). *Pseudomonads* dominate the spoilage flora of pasteurized, homogenized milk at 4–12 °C, and above 12 °C, the spoilage flora is comprised of only 10–20% pseudomonads and the remainder are mesophiles (Chandler and McMeekin 1989). It is long been generally recognized that bacterial growth to a level of $7.5\text{-}\log_{10}$ CFU/mL of refrigerated milk and dairy products represents the end of shelf life (Griffiths et al. 1984). Shelf life of pasteurized milk is greatly affected by the degree of sanitation and postpasteurization recontamination by pseudomonads (Eneroth et al. 2000), but spoilage due to growth of psychrotrophic *Bacillus* also occurs (Ternström et al. 1993) due to spore survival at pasteurization temperatures. Therefore, microbial quality and temperature control of raw milk destined for pasteurization is of great importance and long-term impact on pasteurized products.

Traditional extended shelf-life (ESL) technology in North America incorporates a high heat treatment of the product, which provides normal pasteurized product sensory characteristics, combined with ultraclean packaging, which includes a controlled filling environment and container sterilization (Henyon 1999). This results in product code dates for ESL dairy products in North America ranging from 45 to 60 days, and currently, the technology is expanding to include value-added products along with regular drinking milks. ESL milk has become synonymous for a total system process that encompasses everything from raw product quality to final distribution. Generally, ESL milk has a shelf life longer than pasteurized milk. For ESL milks, particularly if distributed at temperatures higher than 7 °C, the numbers of thermoduric and aerobic sporeformers indicate spoilage. At 8–10 °C, aerobic sporeformers have increased influence, and higher frequencies of psychrotrophic strains are reported. A thermoduric count less than 1000 microorganisms/mL and an aerobic spore count less than 10 microorganisms/mL are typical thresholds for microbial quality.

Butter is microbiologically more stable than other dairy-based products. Pasteurization destroys microorganisms important in butter spoilage, but a variety of these spoilage microorganisms may contaminate the product at some later stage in production (Ayres et al. 1980). Proteolytic and lipolytic microorganisms can bring about flavor defects in refrigerated and ambient stored butter. *Pseudomonas fluorescens* can hydrolyze milk fat, liberating short-chain fatty acids that lead to rancidity. Other bacteria like *Flavobacterium maloloris* and lactic acid bacteria can cause surface taint, leading to putrid, decomposed, and cheesy flavor (Van Zijl and Klapwijk 2000). Spoilage defects have also been caused by yeasts such as *Candida* spp. and molds such as *Aspergillus* spp. Although butter-based products have low a_w, microorganisms can migrate to water droplets (usually <10 μm in size), and there become metabolically active. In rare instances, bacterial pathogens like *Listeria monocytogenes* and *Salmonella* have been isolated from butters and vegetable oil products (Van Zijl and Klapwijk 2000).

Cheeses are also susceptible to microbial growth during shelf life. Microbial growth on hard cheeses is restricted to molds and some yeasts due to an intermediate a_w. Soft and semisoft cheeses are more susceptible to growth by yeast and molds, in particular, and must be prepared and processed in high-care processing environments. Sliced, diced, and shredded cheeses were once limited to a shorter shelf life due to eventual mold growth, until the widespread use of natamycin. Some yogurts, cream cheeses, sour cream, and spreads are preserved with potassium sorbate to inhibit mold and yeast growth and extend shelf life. Fermented products like sour cream and yogurt contain organic acids and sustain a reduced pH that limits microbiological growth to lactic acid bacteria and yeast (Mataragas et al. 2011). Despite high prevalence of organic acids in these products, they can succumb to fungal spoilage before the end of shelf life.

3.7.3 Plant-Based Protein Products

Plant-based protein products meant to be analogous to meat products, like hamburgers, sausages, and roasts, contain proteins from sources such as wheat, soy, pea, and lipids from canola, safflower, and avocado. These products are complex matrices of proteins, lipids, starches, stabilizers, and seasonings designed to achieve the texture and flavor like animal meat products. Due to a high proportion of seasonings and other water-binding texture enhancers, the resulting matrix of such products has a lower pH and a_w than meat products from which they are modeled, thereby inhibiting microbial growth during refrigerated shelf life. Many of these products are refrigerated, vacuum packaged, and assigned a shelf life comparable to meat products, even though the pH and a_w combination may impart additional microbial hurdles compared to meat.

Nut- or legume-based alternatives to dairy, such as almond milk, cashew milk, and soymilk, contain biochemical substrates that are supportive of microbial growth. Just like in animal-derived milks, high-temperature short-time (HTST) pasteurization of nut milk and soymilk reduces vegetative bacteria, yeast, and mold, but allows sporeforming bacteria to survive. Typically, *Bacillus* and *Paenibacillus* dominate pasteurized nut milks, and prolonged, slow cooling after pasteurization can enable germination and proliferation and negatively impact shelf life. Like dairy-based milks, plant-based milks are also susceptible to postprocessing recontamination and will readily spoil if re-contaminated by *Pseudomonas*, lactic acid bacteria, yeast, or mold. Fermented nut milk or soymilk products such as yogurts, sour cream, and cream cheese contain with reduced pH can also succumb to microbial growth either via postprocessing recontamination, failed fermentation, or due to survival and growth of sporeforming bacteria.

3.7.4   Fresh Seafood

The rate of spoilage of fresh seafood is dependent on the number and the type of bacteria, primarily *Enterobacteriaceae* and pseudomonads (Ronsivalli and Charm 1975). Temperature of storage affects the rate of spoilage reactions involving bacterial and autolytic enzymes and affects the rate of bacterial multiplication. The generation time for *Pseudomonas fragi*, a common fish-spoilage bacterium, is about 12 h at 0 °C and only about 2 h at 12.7 °C (Duncan and Nickerson 1961). Spencer and Baines (1964) reported that the spoilage rate of white fish is related approximately linearly to the temperature at which the fish are stored in the range 1–25 °C. In the range of 0–7.8 °C, the rates at which fish spoiled appeared to be linearly related to the temperature at which they were stored (Ronsivalli and Charm 1975). *Pseudomonas* spp. was found to be a major spoilage organism of raw Atlantic Salmon fillets, and H_2S-producing bacteria and *Brochothrix thermosphacta* were also important part of the spoilage microflora (Mikš-Krajnik et al. 2016). Temperature is a significant factor in the types of microorganisms that spoil fresh shrimp, with H_2S-producing bacteria dominating the spoilage microflora at 28 °C and 7 °C and *Pseudomonas* spp. dominating at 0 °C (Dabadé et al. 2015).

As bacteria degrade proteins, the accumulation of ammonia and peptides leads to off odors and flavors. Modified atmosphere packaging of fresh fish can prolong shelf life compared to vacuum packaging. At 0 °C, cod fillets stored under 48% CO_2 extended shelf life by 6–7 days compared to vacuum-packaged fillets, whereas an atmosphere of pure CO_2 only extended shelf life by 2–3 days (Dalgaard et al. 1993). CO_2-rich atmospheric conditions at 0 °C may enable microorganisms such as *Photobacterium phosphoreum* to gain a selective advantage over the H_2S-producing bacteria like *Shewanella putrefaciens*.

Trimethylamine oxide (TMAO) is an important compound for maintenance of the physiological functions in fish and shellfish, but it is also a key substance in the spoilage of raw or processed seafood (Sotelo and Rehbein 2000). TMAO is present at varying levels in fish and shellfish depending on factors such as diet, the age and size of the fish, and environmental factors. Reduction of TMAO into trimethylamine (TMA) is mainly responsible for the off-odor indicative of spoilage and has a very low-odor threshold compared to ammonia. TMAO is reduced to TMA by the action of trimethylamine oxide reductase enzyme, which is possessed by many bacteria within *Enterobacteriaceae* such as *Shewanella* and by *Alteromonas*, *Photobacterium*, and *Vibrio*. TMAO serves as a terminal electron acceptor in anaerobic growth. Many types of bacteria found in spoiling fish are unable to reduce TMAO, and therefore, the usefulness of TMA as a quality parameter is dependent on the storage conditions of the product and the composition of the bacterial flora.

3.7.5 Raw Meat and Poultry

When refrigerated, raw ground poultry in reduced oxygen atmosphere packaging becomes spoiled, microbial counts are at typically $\geq 10^7$ CFU/g (O'Brien and Marshall 1996). However, ground turkey products typically have an order of magnitude higher plate counts than chicken, often reaching 10^8 CFU/g without spoilage (Guthertz et al. 1976). Microorganisms responsible for spoilage of poultry can multiply at refrigeration temperatures (e.g., 1.1–4.4 °C) (International Commission on Microbiological Specifications for Foods [ICMSF] 1986). A criterion for APC in raw chicken of $n = 5$, $c = 3$, $m = 5 \times 10^5$, $M = 10^7$ has long been used, but the ICMSF cautioned that it is not necessarily achievable for turkey even produced in modern processing systems due to the differing microbial load and substrate differences. The ICMSF provides no strict limits to the general microbial population (such as APC or lactic) on raw comminuted meats, but rather suggest that samples of meat at various stages of processing can be used to establish a baseline and understand changes in the microbial population during processing (ICMSF 2011). Despite this, the Canadian Government set guidelines that no sample of ground meat should exceed 7 log CFU/g for aerobic mesophilic counts (Canadian Food Inspection Agency 1999). Culture-dependent and culture-independent characterization of bacterial communities of raw pork stored under modified atmosphere at 4 °C revealed that *Brochothrix thermosphacta*, lactic acid bacteria (*Carnobacterium*, *Lactobacillus*, *Lactococcus*, *Leuconostoc*, and *Weissella*), and *Photobacterium* spp. predominated in pork samples (Nieminen et al. 2016).

Yeasts occur in low numbers on freshly slaughtered cuts of poultry, but can proliferate in minced or ground meats, and can reach 10^6 to 10^7 CFU/g. The yeasts most frequently isolated from comminuted meats are *Candida zeylanoides*, *Candida famata* (*Debaryomyces hansenii*), *Candida sake*, *Yarrowia lipolytica*, *Rhodotorula*, and *Cryptococcus laurentii* (Pitt and Hocking 2009; Viljoen et al. 1998). Often, growth by yeasts leads to the types of quality defects described as doughy or cardboard odor and discoloration.

Modified atmosphere packaging of comminuted meat and poultry is commonly used to extend shelf life and preserve color. Gas flush target compositions can range from 100% N_2 to 70% N_2 with 30% CO_2, to 100% CO_2. Gas compositions with carbon monoxide (CO) are sometimes utilized to fix heme pigments, thereby stabilizing meat and poultry color. For example, a composition of 0.5% CO, 80% CO_2, and 19.5% N_2 for ground poultry will fix the pigment to carboxymyoglobin, which will remain for the duration of the nearly 3-week shelf life. The absence of O_2 will restrict bacterial growth to only facultative anaerobes or anaerobes, not the aerobes like *Pseudomonas*, *Brochothrix*, or *Hafnia*. So, the microflora will shift toward lactic acid bacteria, which will eventually sour the product (due to lactic and acetic acid production). Human consumption of meat that has been packaged in a CO mixture will result in only negligible levels of carboxyhemoglobin in the blood, and it is highly improbable that the use of CO in the packaging of meat will present a toxic threat to consumers (Sørheim et al. 1997). Technologically, there are few drawbacks to the use of CO to form carboxymyoglobin. Some results indicated that carboxymyoglobin is susceptible to lipid-oxidation-induced browning in a pH- and temperature-dependent manner in vitro (not in meat) (Suman et al. 2006).

3.7.6 Further Processed Meat, Poultry, and Fish

Processing imparts changes to the substrate for microorganisms, which basically prolongs the potential shelf life of food. Salting, smoking, and cooking reduce available moisture, alter proteins and fats, and cause lethality to the more sensitive portion of the microbial population. For instance, the shelf life of lightly processed seafood stored at 4 °C is dependent on initial pH and secondarily the NaCl content and the microflora is dominated by lactic acid bacteria and yeasts (Boziaris et al. 2013). With prolonged shelf life comes the need to assure prevention of growth of psychrotrophic bacterial pathogens, such as nonproteolytic *Clostridium botulinum*, *Listeria monocytogenes*, *Bacillus cereus*, and *Yersinia enterocolitica* (covered in Chap. 2). The pathogen of greatest concern on postlethality exposed ready-to-eat (RTE) meats is *Listeria monocytogenes*. Cured RTE meat products, such as ham, can either be formulated with antimicrobial agents to restrict growth of this human pathogen during refrigerated shelf life, or products can be treated with surface application of an antimicrobial agent. Surface application of antimicrobial agents (i.e., postlethality treatments) can have immediate efficacy, resulting in a logarithmic reduction of *L. monocytogenes*, and may have a sustained, inhibitory effect preventing growth throughout shelf life.

Microbial counts have long been used to measure general wholesomeness and potential quality perception of meat products. Spoilage microorganisms are commercially significant in meat products when their numbers reach around 10^7 CFU/g, resulting in sensory changes limiting acceptability and shelf life (Holley et al. 2004; Ingram and Dainty 1971; Mano et al. 1995).

The way processed ready-to-eat meats are formulated, packaged, and stored inherently selects for the slow growth of lactic acid bacteria, which typically do not impart negative sensory changes, and suppresses the growth of microorganisms that can cause unpleasant sensory changes. For example, lactic acid bacteria cause the normally acceptable souring of ham and other lightly cured meat products (Taormina 2014). Lactic acid bacteria are selected by levels of salt, nitrite, organic salts, and other inhibitors, and by vacuum or modified-atmosphere packaging. During long-term refrigerated storage, the lactic acid bacteria metabolize carbohydrates to acidic products, thereby causing a drop in pH and accumulation of organic acids, chiefly lactic acid. The production of acid and bacteriocins by certain species of lactic acid bacteria prevents putrid spoilage, resulting from the ammoniacal products of microbial protein and peptide degradation.

Lactic acid bacteria are ubiquitous and may be introduced to product via the environment, personnel, and processing aids such as brine or lubricants like water, and possibly through raw materials. For example, *Weissella viridescens* is a common spoilage bacterium on cured ham (Karpíšková et al. 2013). This bacterium, formerly known as *Lactobacillus viridescens*, is in the order *Lactobacillales*, family *Leuconostocaceae*, and causes known defects on cured ham such as slime formation and greening. Control of this specific microorganism on RTE meat products would be required to attain shelf-life targets. Once present in a package, bacteria are competing for nutrients and opportunity to increase in population. Certain lactic acid bacteria can hydrolyze sucrose and utilize an enzyme called dextran sucrose, to build a large molecule called dextran. Dextran is the ropy, slimy material seen in packages of cured and uncured luncheon meats and frankfurters. Figure 3.1 depicts the biosynthesis of dextran. The enzyme dextran sucrase is involved in the polymerization reaction, and the energy needed for polymerization comes from the hydrolysis of sucrose (De Vuyst and Degeest 1999).

Although there is much to be explored with regard to the interaction of microbial metabolism, chemical reactions, and bearing on shelf life at different temperatures, some progress has been made to understand the microbiological and physiochemical changes occurring in a glucose-supplemented broth system with three *Lactobacillus sakei* strains, two *Lactobacillus curvatus* strains, two *Lactobacillus plantarum* strains, and two *Lactobacillus paracasei* strains (Vaikousi et al. 2008). There was a temperature-dependent impact on the ultimate levels of lactic acid and viable counts of lactic acid bacteria. After 450 h at 4 °C, lactic acid bacteria reached 8 $\log_{10}$ CFU/mL, pH declined from 6 down to 4, and lactic acid levels rose to

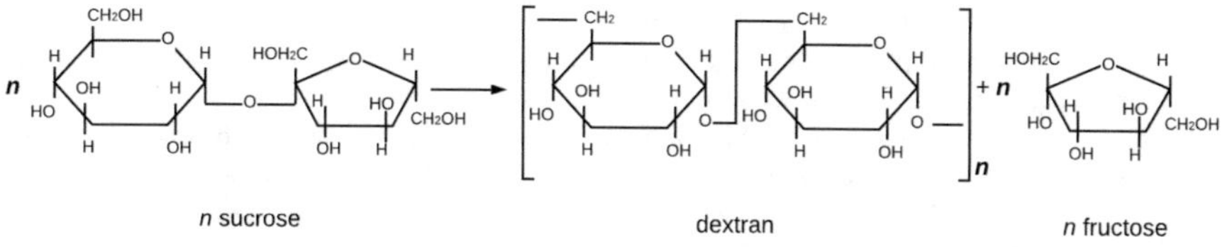

Fig. 3.1 Biosynthesis of the homopolysaccharide dextran

8 mM. After 200 h at 8 °C, lactic acid bacteria reached $8 \log_{10}$ CFU/mL, pH declined from 6 down to 4, and lactic acid levels rose to nearly 10 mM. At 12 °C, similar results were seen with plate counts and pH within 180 h, but the lactic acid levels rose to 16 mM. These data indicate that temperature obviously has a bearing on rate of microbial growth, but also upon the levels of lactic acid produced in the medium. This could provide clues as to how food spoilage occurs and manifests differently at different temperatures. The same microorganisms may be present, the same populations are ultimately reached, but abusive temperature would impart a much different biochemical profile than slower growth at refrigeration temperature.

3.8   Conclusion

Microbial growth is often but not always the primary determinant of shelf life. Temperature is often the primary determinant of microbial growth. Several microorganisms have minimum growth temperatures at or below the storage temperatures of perishable foods and tend to outcompete other microorganisms and predominate at the end of shelf life. In some food products processed in certain ways, innocuous growth of homofermentative lactic acid bacteria does not correlate with sensory loss. Standard plate count techniques are widely used for day-to-day shelf-life testing, although more rapid and precise techniques are available and are used in research and special projects within industry. Techniques that identify chemical byproducts of microbial metabolism can serve as early warning of end of shelf life of foods. Spoilage microflora varies with different food commodities, but certain psychrotrophic bacteria and molds are comprised of the spoilage flora of multiple categories of foods. Processing techniques, food ingredients, and packaging shift eliminate or inhibit those microorganisms that have the greatest capability of spoilage, leading to extending shelf life of foods.

References

Amodio, M.L., A. Derossi, L. Mastrandrea, G.B. Martínez Hernández, and G. Colelli. 2015. The use of multivariate analysis as a method for obtaining a more reliable shelf-life estimation of fresh-cut produce: A study on pineapple. In III international conference on fresh-cut produce: Maintaining quality and safety. *Acta Horticulturae* 1141: 131–136.

Ayres, J.C., J.O. Mundt, and W.E. Sandine. 1980. *Microbiology of Foods, Chp 16—Dairy Products*. San Francisco: W.H. Freeman and Company.

Baranyi, J., and T.A. Roberts. 1994. A dynamic approach to predicting bacterial growth in food. *International Journal of Food Microbiology* 23 (3–4): 277–294.

Boziaris, I.S., A.P. Stamatiou, and G.J.E. Nychas. 2013. Microbiological aspects and shelf life of processed seafood products. *Journal of the Science of Food and Agriculture* 93 (5): 1184–1190.

Brackett, R.E. 1987. Microbiological consequences of minimally processed fruits and vegetables. *Journal of Food Quality* 10 (3): 195–206.

Brooks, F.T. and C.G. Hansford. 1923. Mould growths upon cold-store meat. Transactions of the British Mycological Society 8(3): 113–142.

Canadian Food Inspection Agency. 1999. Sampling and testing procedures. In *In Meat Hygiene Manual, edited by Meat and Animal Product Division, 18*. Canada: Government of Canada.

Chandler, R.E., and T.A. McMeekin. 1989. Temperature function integration as the basis of an accelerated method to predict the shelf life of pasteurized, homogenized milk. *Food Microbiology* 6 (2): 105–111.

Condurso, C., A. Verzera, V. Romeo, M. Ziino, and F. Conte. 2008. Solid-phase microextraction and gas chromatography mass spectrometry analysis of dairy product volatiles for the determination of shelf-life. *International Dairy Journal* 18 (8): 819–825.

Corradini, M.G., and M. Peleg. 2007. Shelf-life estimation from accelerated storage data. *Trends in Food Science & Technology* 18: 37–47.

D'Aoust, J.Y., 1991. Psychrotrophy and foodborne Salmonella. International Journal of Food Microbiology 13(3): 207–215.

Dabadé, D.S., H.M. den Besten, P. Azokpota, M.R. Nout, D.J. Hounhouigan, and M.H. Zwietering. 2015. Spoilage evaluation, shelf-life prediction, and potential spoilage organisms of tropical brackish water shrimp (*Penaeus notialis*) at different storage temperatures. *Food Microbiology* 48: 8–16.

Dalgaard, P., L. Gram, and H.H. Huss. 1993. Spoilage and shelf-life of cod fillets packed in vacuum or modified atmospheres. *International Journal of Food Microbiology* 19 (4): 283–294.

De Vuyst, L., and B. Degeest. 1999. Heteropolysaccharides from lactic acid bacteria. *FEMS Microbiology Reviews* 23 (2): 153–177.

Derossi, A., L. Mastrandrea, M.L. Amodio, M.L.V. de Chiara, and G. Colelli. 2016. Application of multivariate accelerated test for the shelf life estimation of fresh-cut lettuce. *Journal of Food Engineering* 169: 122–130.

Domsch, K.H., Gams, W. and Anderson, T.H. 1980. Compendium of soil fungi. Volume 1. Academic Press (London) Ltd.

Duncan, D. W., Jr. , and J. T. R. Nickerson. 1961. Effect or environmental and physiological conditions on the exponential phase growth of Pseudomonas fragi (ATCC 4973). In Proc. Low Temp. Microbiol. Symp., p. 253-262. Campbell Soup Co., Camden , N.J. Panasenko, V.T., 1967. Ecology of microfungi. The Botanical Review 33(3): 189–215.

El-Gendy, S.M., H. Abdel-Galil, Y. Shahin, and F.Z. Hegazi. 1983. Acetoin and diacetyl production by homo-and heterofermentative lactic acid bacteria. *Journal of Food Protection* 46 (5): 420–425.

Eneroth, Å., S. Ahrné, and G. Molin. 2000. Contamination routes of Gram-negative spoilage bacteria in the production of pasteurised milk, evaluated by randomly amplified polymorphic DNA (RAPD). *International Dairy Journal* 10 (5–6): 325–331.

Farber, J.M. 1991. Microbiological aspects of modified-atmosphere packaging technology—A review. *Journal of Food Protection* 54: 58–70.

Gougouli, M., and K.P. Koutsoumanis. 2010. Modelling growth of *Penicillium expansum* and *Aspergillus niger* at constant and fluctuating temperature conditions. *International Journal of Food Microbiology* 140 (2–3): 254–262.

Gram, L., L. Ravn, M. Rasch, J.B. Bruhn, A.B. Christensen, and M. Givskov. 2002. Food spoilage—Interactions between food spoilage bacteria. *International Journal of Food Microbiology* 78 (1–2): 79–97.

Griffiths, M.W., J.D. Phillips, and D.D. Muir. 1984. Methods for rapid detection of post-pasteurization contamination in cream. *International Journal of Dairy Technology* 37 (1): 22–26.

Guthertz, Linda S., John T. Fruin, Delano Spicer, and James L. Fowler. 1976. Microbiology of fresh comminuted Turkey meat. *Journal of Milk and Food Technology* 39 (12): 823–829.

Guy, R.A., A. Kapoor, J. Holicka, D. Shepherd, and P.A. Horgen. 2006. A rapid molecular-based assay for direct quantification of viable bacteria in slaughterhouses. *Journal of Food Protection* 69 (6): 1265–1272.

Henyon, D.K. 1999. Extended shelf-life milks in North America: A perspective. *International Journal of Dairy Technology* 52 (3): 95–101.

Holley, R.A. 1999. Brochothrix. In *Encyclopedia of food microbiology*, ed. R.K. Robinson, C.A. Batt, and P. Patel, 314–318. New York: Academic.

Holley, R.A., M.D. Peirson, J. Lam, and K.B. Tan. 2004. Microbial profiles of commercial, vacuum-packaged, fresh pork of normal or short storage life. *International Journal of Food Microbiology* 97 (1): 53–62.

Hug, D.H., and J.K. Hunter. 1974. Effect of temperature on histidine ammonia-lyase from a psychrophile, *Pseudomonas putida*. *Journal of Bacteriology* 119 (1): 92–97.

in't Veld, J.H.H. 1996. Microbial and biochemical spoilage of foods: An overview. *International Journal of Food Microbiology* 33 (1): 1–18.

Ingram, M., and R.H. Dainty. 1971. Changes caused by microbes in spoilage of meats. *Journal of Applied Bacteriology* 34 (1): 21–39.

International Commission on Microbiological Specifications for Foods. 1986. *Microorganisms in foods 2. Sampling for microbiological analysis: Principles and specific applications*. 2nd ed. Toronto: University of Toronto Press.

———. 1996. *Microorganisms in Foods 5: Characteristics of Microbial Pathogens*. Chapman & Hall: Blackie Academic & Professional.

———. 2011. *Microorganisms in Foods*: 8. https://doi.org/10.1007/978-1-4419-9374-8.

Ioannidis, A.G., F.M. Kerckhof, Y.R. Drif, M. Vanderroost, N. Boon, P. Ragaert, B. De Meulenaer, and F. Devlieghere. 2018. Characterization of spoilage markers in modified atmosphere packaged iceberg lettuce. *International Journal of Food Microbiology* 279: 1–13.

Jacxsens, L., F. Devlieghere, and J. Debevere. 2002. Temperature dependence of shelf-life as affected by microbial proliferation and sensory quality of equilibrium modified atmosphere packaged fresh produce. *Postharvest Biology and Technology* 26 (1): 59–73.

Janssen, M.M.T., H.M.C. Put, and M.J.R. Nout. 1996. Natural toxins. In *Food Safety and Toxicity*, ed. J. de Vries, 7–37. Open University of the Netherlands: Publication Robert B. Stern.

Karpíšková, R., M. Dušková, and J. Kameník. 2013. *Weissella viridescens* in meat products—a review. *Acta Veterinaria Brno [serial online]*. 82 (3): 237–241.

King, A.D., and H.R. Bolin. 1989. Physiological and microbiological storage stability of minimally processed fruits and vegetables. *Food Technology* 43 (2): 132–135.

Korkeala, H.J., P.M. Makela, and H.L. Suominen. 1990. Growth temperatures of ropy slime-producing lactic acid bacteria. *Journal of Food Protection* 53 (9): 793–794.

Koutsoumanis, K. 2001. Predictive modeling of the shelf life of fish under nonisothermal conditions. *Applied and Environmental Microbiology* 67 (4): 1821–1829. https://doi.org/10.1128/AEM.67.4.1821-1829.2001.

Kraft, A. 1992. *Psychrotrophic Bacteria in Foods: Disease and Spoilage*. Boca Raton, FL: CRC Press Inc.

Labuza, T.P. 1984. Application of chemical kinetics to deterioration of foods. *Journal of Chemical Education* 61: 348.

Leroi, F., P.A. Fall, M.F. Pilet, F. Chevalier, and R. Baron. 2012. Influence of temperature, pH and NaCl concentration on the maximal growth rate of *Brochothrix thermosphacta* and a bioprotective bacteria *Lactococcus piscium* CNCM I-4031. *Food Microbiology* 31: 222–228.

Lin, M., M. Al-Holy, M. Mousavi-Hesary, H. Al-Qadiri, A.G. Cavinato, and B.A. Rasco. 2004. Rapid and quantitative detection of the microbial spoilage in chicken meat by diffuse reflectance spectroscopy (600–1100 nm). *Letters in Applied Microbiology* 39 (2): 148–155.

Longhi, D.A., A. Tremarin, B.A.M. Carciofi, J.B. Laurindo, and G.M.F. de Aragão. 2014. Modeling the growth of *Byssochlamys fulva* on solidified apple juice at different temperatures. *Brazilian Archives of Biology and Technology* 57 (6): 971–978.

Longhi, D.A., W.F. Martins, N.B. da Silva, B.A.M. Carciofi, G.M.F. de Aragão, and J.B. Laurindo. 2016. Estimation of the temperature dependent growth parameters of *Lactobacillus viridescens* in culture medium with two-step modelling and optimal experimental design approaches. *Procedia Food Science* 7: 25–28.

Lund, B.M., T.C. Baird-Parker, and G.W. Gould. 2000. *The Microbiological Safety and Quality of Food*. Vol. 1. Gaithersburg, MD: Aspen Publishers, Inc.

Lyhs, U., J.M. Koort, H.S. Lundström, and K.J. Björkroth. 2004. *Leuconostoc gelidum* and *Leuconostoc gasicomitatum* strains dominated the lactic acid bacterium population associated with strong slime formation in an acetic-acid herring preserve. *International Journal of Food Microbiology* 90 (2): 207–218.

Mackelprang, R., A. Burkert, M. Haw, T. Mahendrarajah, C.H. Conaway, T.A. Douglas, and M.P. Waldrop. 2017. Microbial survival strategies in ancient permafrost: Insights from metagenomics. *The ISME Journal* 11 (10): 2305.

Man, D. 2002. *Shelf Life*. London: Blackwell Science Ltd.

Mano, S.B., G.D. Garcia de Fernando, D. Lopez-Galvez, M.D. Selgas, M.L. Garcia, M.I. Cambero, and J.A. Ordonez. 1995. Growth/survival of natural flora and *Listeria monocytogenes* on refrigerated uncooked pork and Turkey packaged under modified atmospheres. *Journal of Food Safety* 15: 305–319.

Martínez, N., M.C. Martín, A. Herrero, M. Fernández, M.A. Alvarez, and V. Ladero. 2011. qPCR as a powerful tool for microbial food spoilage quantification: Significance for food quality. *Trends in Food Science & Technology* 22 (7): 367–376.

Mataragas, M., V. Dimitriou, P.N. Skandamis, and E.H. Drosinos. 2011. Quantifying the spoilage and shelf-life of yoghurt with fruits. *Food Microbiology* 28 (3): 611–616.

Mikš-Krajnik, M., Y.J. Yoon, D.O. Ukuku, and H.G. Yuk. 2016. Volatile chemical spoilage indexes of raw Atlantic salmon (*Salmo salar*) stored under aerobic condition in relation to microbiological and sensory shelf lives. *Food Microbiology* 53: 182–191.

Mizrahi, S. 2004. Accelerated shelf life tests. In *Understanding and Measuring the Shelf-Life Of Food*, ed. R. Steele. Cambridge, UK: Woodhead Publishing.

Mislivec, P.B. and Tuite, J., 1970. Temperature and relative humidity requirements of species of Penicillium isolated from yellow dent corn kernels. Mycologia 62(1): 75–88.

Mossel, D.A.A., and M. Ingram. 1955. The physiology of the microbial spoilage of foods. *Journal of Applied Bacteriology* 18 (2): 232–268.

Nieminen, T.T., P. Dalgaard, and J. Björkroth. 2016. Volatile organic compounds and *Photobacterium phosphoreum* associated with spoilage of modified-atmosphere-packaged raw pork. *International Journal of Food Microbiology* 218: 86–95.

O'Brien, J.K., and Robert T. Marshall. 1996. Microbiological quality of raw ground chicken processed at high isostatic pressure. *Journal of Food Protection* 59 (2): 146–150.

O'Connor-Shaw, R.E., R. Roberts, A.L. Ford, and S.M. Nottingham. 1994. Shelf life of minimally processed honeydew, kiwifruit, papaya, pineapple and cantaloupe. *Journal of Food Science* 59 (6): 1202–1206.

Pablo, B.D., M.A. Asensio, B. Sanz, and J.A. Ordonez. 1989. The D (−) lactic acid and acetoin/diacetyl as potential indicators of the microbial quality of vacuum-packed pork and meat products. *Journal of Applied Bacteriology* 66 (3): 185–190.

Panagou, E.Z., P.N. Skandamis, and G.-J.E. Nychas. 2003. Modelling the combined effect of temperature, pH and a_w on the growth rate of *Monascus ruber*, a heat-resistant fungus isolated from green table olives. *Journal of Applied Microbiology* 94 (1): 146–156.

Panagou, E.Z., S. Chelonas, I. Chatzipavlidis, and G.-J.E. Nychas. 2010. Modelling the effect of temperature and water activity on the growth rate and growth/no growth interface of *Byssochlamys fulva* and *Byssochlamys nivea*. *Food Microbiology* 27 (5): 618–627.

Park, J.-M., J.-H. Koh, and J.-M. Kim. 2018. Predicting Shelf-life of Ice Cream by Accelerated Conditions. Korean Journal for Food Science of Animal Sesources 38 (6): 1216–1225.

Peleg, M., M.D. Normand, and M.G. Corradini. 2012. The Arrhenius equation revisited. *Critical Reviews in Food Science and Nutrition* 52 (9): 830–851.

Pereira, J.A., L. Dionísio, L. Patarata, and T.J. Matos. 2019. Multivariate nature of a cooked blood sausage spoilage along aerobic and vacuum package storage. *Food Packaging and Shelf Life* 20: 100304.

Pitt, J.I., and A.D. Hocking. 1997. *Fungi and Food Spoilage*. 2nd ed. Chapman & Hall: Blackie Academic & Professional.

Pitt, John I., and Ailsa D. Hocking. 2009. *Fungi and Food Spoilage*. New York: Springer Science & Business Media.

Pothakos, V., S. Samapundo, and F. Devlieghere. 2012. Total mesophilic counts underestimate in many cases the contamination levels of psychrotrophic lactic acid bacteria (LAB) in chilled-stored food products at the end of their shelf-life. *Food Microbiology* 32 (2): 437–443.

Pothakos, V., C. Snauwaert, P. De Vos, G. Huys, and F. Devlieghere. 2014. Psychrotrophic members of *Leuconostoc gasicomitatum*, *Leuconostoc gelidum* and *Lactococcus piscium* dominate at the end of shelf-life in packaged and chilled-stored food products in Belgium. *Food Microbiology* 39: 61–67.

Price, P.B., and T. Sowers. 2004. Temperature dependence of metabolic rates for microbial growth, maintenance, and survival. *Proceedings of the National Academy of Sciences* 101 (13): 4631–4636.

Ricciardi, A., E. Parente, and T. Zotta. 2009. Modelling the growth of *Weissella cibaria* as a function of fermentation conditions. *Journal of Applied Microbiology* 107 (5): 1528–1535.

Riva, M., D. Fessas, and A. Schiraldi. 2001. Isothermal calorimetry approach to evaluate shelf life of foods. *Thermochimica Acta* 370 (1–2): 73–81.

Ronsivalli, L.J., and S.E. Charm. 1975. Spoilage and shelf life prediction of refrigerated fish. *Marine Fisheries Review* 37 (4): 32–34.

Rouf, M.A., and M.M. Rigney. 1971. Growth temperatures and temperature characteristics of *Aeromonas*. *Applied Microbiology* 22 (4): 503–506.

Rysstad, G., and J. Kolstad. 2006. Extended shelf life milk—Advances in technology. *International Journal of Dairy Technology* 59 (2): 85–96.

Samson, R.A., J. Houbraken, U. Thrane, J.C. Frisvad, and B. Andersen. 2010. *Food and Indoor Fungi*. Utrecht, The Netherlands: CBS-KNAW Fungal Biodiversity Centre.

Saraiva, C., I. Oliveira, J.A. Silva, C. Martins, J. Ventanas, and C. García. 2015. Implementation of multivariate techniques for the selection of volatile compounds as indicators of sensory quality of raw beef. *Journal of Food Science and Technology* 52 (6): 3887–3898.

Sørheim, O., T. Aune, and T. Nesbakken. 1997. Technological, hygienic and toxicological aspects of carbon monoxide used in modified-atmosphere packaging of meat. *Trends in Food Science & Technology* 8 (9): 307–312.

Sotelo, C.G., and H. Rehbein. 2000. *TMAO-Degrading Enzymes. Food Science and Technology*, 167–190. New York: Marcel Dekker.

Spencer, R. and C. R. Baines. 1964. The effect or temperature on the spoilage of wet white fish. Food Technology 18:769–773.

Stannard, C.J., A.P. Williams, and P.A. Gibbs. 1985. Temperature/growth relationships for psychrotrophic food-spoilage bacteria. *Food Microbiology* 2 (2): 115–122.

Suman, S.P., R.A. Mancini, and C. Faustman. 2006. Lipid-oxidation-induced carboxymyoglobin oxidation. *Journal of Agricultural and Food Chemistry* 54 (24): 9248–9253.

Szabo, E.A., K.J. Scurrah, and J.M. Burrows. 2000. Survey for psychrotrophic bacterial pathogens in minimally processed lettuce. Letters in Applied Microbiology 30(6), 456–460.

Taiti, C., C. Costa, P. Menesatti, S. Caparrotta, N. Bazihizina, E. Azzarello, W.A. Petrucci, E. Masi, and E. Giordani. 2015. Use of volatile organic compounds and physicochemical parameters for monitoring the post-harvest ripening of imported tropical fruits. *European Food Research and Technology* 241 (1): 91–102.

Taormina, P.J. 2014. Meat and poultry: Curing of meat. *Encyclopedia of Food Microbiology (Second Edition)* 2: 501–507.

Ternström, A., A.M. Lindberg, and G. Molin. 1993. Classification of the spoilage flora of raw and pasteurized bovine milk, with special reference to *Pseudomonas* and *Bacillus*. *Journal of Applied Bacteriology* 75 (1): 25–34.

Tománková, J., J. Bořilová, I. Steinhauserová, and L. Gallas. 2012. Volatile organic compounds as biomarkers of the freshness of poultry meat packaged in a modified atmosphere. *Czech Journal of Food Sciences* 30 (5): 395–403.

Vaikousi, H., C.G. Biliaderis, and K.P. Koutsoumanis. 2008. Development of a microbial time/temperature indicator prototype for monitoring the microbiological quality of chilled foods. *Applied and Environmental Microbiology* 74 (10): 3242–3250.

Van Zijl, M.M., and P.M. Klapwijk. 2000. Yellow fat products (butter, margarine, dairy and non-dairy spreads), Chp 29. In *The Microbiological Safety and Quality of Food*, ed. B.M. Lund, T.C. Baird-Parker, and G.W. Gould, vol. I. Gaithersburg, MD: Aspen.

Viljoen, B.C., I. Geornaras, A. Lamprecht, and A. Von Holy. 1998. Yeast populations associated with processed poultry. *Food Microbiology* 15 (1): 113–117.

Wijtzes, T., J.C. De Wit, J.H.J. in't Veld, K. Van't Reit, and M.H. Zwietering. 1995. Modelling bacterial growth of *Lactobacillus curvatus* as a function of acidity and temperature. *Applied and Environmental Microbiology* 61 (7): 2533–2539.

Wilson, P.D.G., T.F. Brocklehurst, S. Arino, D. Thuault, M. Jakobsen, M. Lange, J. Farkas, J.W.T. Wimpenny, and J.F. Van Impe. 2002. Modelling microbial growth in structured foods: Towards a unified approach. *International Journal of Food Microbiology* 73 (2–3): 275–289.

Xiao, Z., and J.R. Lu. 2014. Strategies for enhancing fermentative production of acetoin: A review. *Biotechnology Advances* 32 (2): 492–503.

Yamasato, K., D. Okuno, and T. Ohtomo. 1973. Preservation of bacteria by freezing at moderately low temperatures. *Cryobiology* 10: 453–463.

Zeppa, G., L. Conterno, and V. Gerbi. 2001. Determination of organic acids, sugars, diacetyl, and acetoin in cheese by high-performance liquid chromatography. *Journal of Agricultural and Food Chemistry* 49 (6): 2722–2726.

Zhang, Y., Y. Mao, K. Li, P. Dong, R. Liang, and X. Luo. 2011. Models of *Pseudomonas* growth kinetics and shelf life in chilled *Longissimus dorsi* muscles of beef. *Asian-Australasian Journal of Animal Sciences* 24 (5): 713–722. https://doi.org/10.5713/ajas.2011.10404.

Zwietering, M.H., J.C. De Wit, H.G.A.M. Cuppers, and K. Vann't Reit. 1994. Modeling of bacterial growth with shifts in temperature. *Applied and Environmental Microbiology* 60 (1): 204–213.

Chapter 4
Impact of Sanitation on Product Shelf Life

Steven Tsuyuki and Margaret D. Hardin

4.1 Introduction

Maximizing and maintaining consistent product shelf life continues to challenge the food industry. With the health implications (Doyle and Glass 2010; Taormina 2010) of sodium and the consumers' desire for products with a "clean" label, food product susceptibility to microbial spoilage and food safety risks has increased. At the same time, the economics of food manufacturing has shifted production toward mechanization and automation. Longer production runs utilizing complex equipment with shorter, more intense sanitation cycles have become the norm. Within the healthy tension that exists between running the plant to meet customer orders and cleaning the plant, production will always be "king." However, the consequence of not having enough importance placed on sanitation can lead to the disruption of production activities due to regulatory enforcement and/or food safety and quality product issues, and result in product recalls and loss of customer and consumer confidence. A new paradigm is proposed: one that sees sanitation as supporting production and providing the foundation for the manufacture of safe finished goods that consistently meet the stated shelf life. In this paradigm, the new production day starts with a clean plant.

Under practical conditions, the manufacturing facility cannot precisely control all types of microorganisms found in finished goods, but the plant can control the number of microbes and the risks associated with product contamination. If the microbial

S. Tsuyuki (✉)
Sanitary Design and Corporate Sanitation, Maple Leaf Foods, Mississauga, ON, Canada
e-mail: steven.tsuyuki@mapleleaf.com

M. D. Hardin
Vice President of Technical Services, IEH Laboratories and Consulting Group,
Lake Forest Park, WA, USA
e-mail: mh@iehinc.com

© Springer Nature Switzerland AG 2021
P. J. Taormina, M. D. Hardin (eds.), *Food Safety and Quality-Based Shelf Life of Perishable Foods*, Food Microbiology and Food Safety,
https://doi.org/10.1007/978-3-030-54375-4_4

numbers on or in products are excessive, hurdles such as formulation, packaging conditions, and refrigeration may not prevent the proliferation of pathogens or spoilage organisms to sufficiently assure product safety or quality attributes at the stated code date on the product package. Plant sanitation, therefore, cannot be overemphasized (Allen and Foster 1960). Since the 1980s, the meat industry has developed tactics to control the persistence and spread of *Listeria* in the processing environment (Malley et al. 2015). There is common ground between controlling premature spoilage by spoilage organisms such as lactic acid bacteria (LAB) and controlling bacterial pathogens that are capable of growth at refrigeration temperatures such as *Listeria*. Both of these groups of microorganisms are ubiquitous, environmental microbes that often persist in food-manufacturing environments. The factors that control *Listeria* also form the foundation required to control spoilage organisms.

Traffic Control + GMP + Dry, Uncracked Floors + Sanitation + Sanitary Design = Control

Managing the environment in which post-lethality exposed products interact with people and equipment is crucial to control both *Listeria* and lactics. While cooking can eliminate spoilage organisms such as LAB that may be present in the raw meats, cooked products may become re-contaminated during handling, peeling, slicing, portioning, and packaging and recommendations published in the scientific literature include more frequent cleaning of the packaging equipment (Kempton and Bobier 1970).

This chapter focuses on the sanitation program utilized by successful food processors and provides industry best practices on the execution of the seven-step daily cleaning and sanitation process. In addition, the role of non-daily sanitation tasks is discussed. The importance of equipment sanitary design and installation that promotes accessibility and enables sanitation performance to meet levels of microorganisms necessary to maximize shelf life as well as for pathogen control is discussed. Finally, performance metrics beyond costs are described including environmental swabbing, which often drives continuous improvement in both sanitation effectiveness and efficiency. A maturity model across these attributes is provided as a tool for benchmarking and self-assessment.

4.2   Sanitation Program

An effective sanitation program is key to controlling food safety issues and maintaining product shelf life. The type and frequency of cleaning depends on the complexity of the process, equipment design and accessibility, infrastructure, and types of soils involved. A complete and effective sanitation program consists of both daily sanitation task as well as those tasks that are performed less than daily. Together, daily and non-daily sanitation tasks form the basis of a complete sanitation program and should be captured in the facility's Master Sanitation Schedule (MSS). Specialized tasks may be developed to address a specific piece of equipment and surrounding area or for an infrastructure area such as a product storage cooler. Non-daily tasks such as deep cleaning are intensified cleaning procedures that often involve the extensive, and more time-consuming, disassembly of equipment areas that cannot be completed daily. Production activities may allow food residues and fines from foods, liquids from marinades, pinfeathers, fish scales, seeds, sprouts, and vegetable peelings, to become trapped inside equipment well beyond what is visible on the surface. The extensive disassembly of equipment increases access to surfaces such as cracks, crevices, and pores, which are difficult to reach during daily sanitation, and enables these areas to be manually cleaned and sanitized. Thorough disassembly of equipment can be time-consuming and labor-intensive and requires the coordinated efforts of personnel in production, maintenance, engineering, sanitation, quality assurance (QA), and food safety departments. Sanitors need support from maintenance personnel to break down equipment or access cabinets that house electrical components. Some outside expertise such as a sanitation specialist, equipment manufacturer, refrigeration expert, or the sanitation chemical representative may also be of value. Due to the extensive nature of the disassembly process, deep cleaning often occurs on weekends or on planned prolonged shutdowns. Extensive disassembly of equipment may lead to opening up previously unexplored areas of equipment, and therefore, participation on the part of maintenance personnel to assist in the disassembly process is invaluable. The process can also lead to opportunities to redesign equipment and/or parts to be more easily removed and cleaned in the future. During deep cleaning and disassembly, it is imperative to keep track of parts such as screws, bolts, hoses, wear strips, and guides for equipment such as conveyors. Small parts can be placed in tubs and labeled with the equipment or conveyor number in order to keep track of conveyor belts and associated guides, parts, and pieces. A map of the room and lines with a number or color code can also assist in keeping track of parts and pieces. Following removal, conveyor belts can be put into vats for scrubbing or laid out on tables in order to access all areas for scrubbing and cleaning. Finally, operational sanitation tasks that occur within the production window should be included in the facility's Good Manufacturing Practices (GMPs). Operational sanitation tasks are typically executed by front-line employees or designees to ensure that the last hour of production is just as clean as the first hour.

4.2.1 Daily Sanitation: Seven-Step Process

Most food-processing plants operate one or two shifts of production, followed by a third shift for cleaning and sanitizing. Structuring sanitation as a step-by-step process drives organization, order, and method into what appears to be a chaotic whirlwind of food debris and water, very much like being inside a dishwasher. If sanitors follow a defined process, understand their role, are trained on the performance expectations for each step, are supported for success, are supervised to work as a team, and are held accountable for their performance, then consistent outcomes will be achieved that balance the efficiency of cleaning with consistent sanitation effectiveness.

Support from plant management and supervision are required for the successful implementation of a consistent process. Sanitation too often receives the least amount of support of any function in most plants. It is a tough statement to make, but the reality is that sanitation happens when most people are sleeping. It is often a challenge to have people available to verify the sanitation process, particularly when it occurs in the third shift. It is advisable for the individuals responsible for verification of sanitation performance to be present on Thursday night so that when Friday morning rolls around, the weekend can be used to recover. Sleep deprivation is a small price to pay to show the sanitation team that they are relevant. This is often a teachable moment for the members of QA and operations who are verifying the process, as the challenges the sanitation team may be facing on a day-to-day basis become evident, and hopefully results in all parties working together to facilitate the effectiveness of the sanitation process. In some plants, the sanitation shift is moved to the afternoon shift in order to improve sanitation support. Setting up sanitors for success is pretty simple. A plant that works production to a defined end time, transitions the room for cleaning by removing all production materials from the area, and completes specific tasks assigned to production will enable an efficient start to the sanitation process. Preserving the sanitation window is critical. If production runs late and transition of the room to sanitation is delayed, too often the expectation is that the sanitation window is compressed in order for the next production shift to start up on time. Nothing is more demotivating and stressful than starting sanitation late, receiving the area in an unacceptable state, and rushing to complete the work to be on time for production startup. Things do go wrong on occasion, and equipment breaks and goes down, forcing production to run later than originally planned. In these instances, the amount of time allotted for sanitation must be maintained even if production needs to start later the next day. This must be non-negotiable as the sanitation process is too important to maintain the quality of the work and the safety and the morale of the sanitors. The sanitation shift is not the time to play catch up for production delays and downtime. Strong supervision is the key to consistent process execution. This is even more important for sanitation. First, absenteeism and turnover are often much higher for sanitation than for production. This places a tremendous burden on recruitment and training. Often, short staffing leads to "working" supervisors that are unable to monitor and coach their front-line sanitors to work as a team, and are not available to hold them accountable for their performance. Many companies are realizing the importance of sanitation and are investing in the sanitation process by ensuring that they have proper staffing, including supervisors, adequate tools to perform their jobs, and offering higher salaries than are paid to production employees.

The objectives of the seven-step wet sanitation process are to:

1. Remove loose soils and fat films using pressurized hot water
2. Remove protein and residue build-up/films, and any associated microorganisms using detergent and scrubbing
3. Sanitize surfaces using low-pressure, high-volume chemical application to further reduce microorganisms to acceptable levels

It is very important to complete each process step satisfactorily, in the correct order, and to stay in sequence with other sanitors working in the same area. The seven-step sanitation process and best practice elements are described as follows:

1. Area Preparation/Dry Pickup

   During this stage, the area is in transition from production to sanitation. Production and sanitation employees must work together to complete room transfer tasks. No water is used at this point. Removing gross soils using squeegees, scrapers, and shovels is far more efficient and effective than using water to knock down soils. The amount of dry soils can be captured as a performance metric.

   (a) Stage sanitation equipment, tools, and reusable Personal Protective Equipment (PPE) required for each area/equipment, inspect items for wear/damage, and repair/replace as required.
   (b) Switch air-handling equipment to sanitation mode, if applicable.
   (c) Perform a documented room transfer audit with production and sanitation supervisors to ensure that all production-related tasks are completed. Tasks include the following:

   - Collect and remove all edible and inedible product from the area to be cleaned.
   - Collect and remove all packaging materials and production supplies from the area to be cleaned.
   - Collect and remove all garbage and scrape floors and equipment. Remove all debris into designated disposal areas.
   - Clean, sanitize, and secure all water-sensitive components in the area with plastic to prevent water damage. Some companies clean and sanitize water-sensitive components both before covering (prior to sanitation) and after removal of the cover. The additional step of cleaning after removal of the covering (plastic) also reduces the risk from cross-contamination that may occur during the removal of the cover (plastic) and any food debris, overspray, water, and chemicals that may accumulate on the cover during sanitation.

   (d) Disassemble equipment to the level required for daily sanitation. These tasks are performed by production, maintenance, or sanitors. Sanitors must ensure that all equipment has been de-energized and proper Lock-Out Tag-Out (LOTO) procedures have been performed prior to interacting with equipment.
   (e) Remove remaining visible soils from equipment and surrounding areas using squeegees, scrapers, and shovels.
   (f) Inspect the condition of the equipment for missing parts or pieces, improperly secured or covered equipment, damaged parts, or foreign material residue. During this process, the sanitor is able to evaluate the condition of equipment and see any damaged or worn surfaces and parts that may not be apparent during production until equipment failure occurs. Reporting these conditions is invaluable to the preventive maintenance of equipment and ultimately reduces food safety risks.

2. Pre-Rinse

I have never met a sanitor that upon the first meeting and asking them, "how can I make your job easier?" did not tell me that they needed more water pressure or more consistent temperature for hot water. However, the use of excessive pressurized water increases food safety risk and will degrade equipment assets by compromising seals and forcing debris and water deeper into equipment and enclosures. If water pressure alone is not high enough, sanitors will too often modify the nozzle tips to spray in a "bullet" style manner, which results in maximizing point pressure while minimizing dispersion. In effect, sanitors are cleaning away soils using something akin to a "waterpik." In addition to the inefficiency in using point pressure, the effectiveness of the sanitation process is negatively impacted due to the potential for water and soil to penetrate deeper into complex equipment beyond what is accessible and cleaned daily. Growth niches exposed to water and soils, if inhabited by *Listeria* or spoilage bacteria, can lead to harborage sites that allow bacteria to grow despite daily sanitation. Shedding of the bacteria occurs through machinery motion, food interaction with equipment, and the interaction of the exposed process line with the surrounding infrastructure. Shedding and spreading though transfer vectors result in environmental positives for *Listeria* or high microbial counts on food contact and non-food-contact surfaces that can lead to product contamination during operations. Proper training, supervision, and coaching will help sanitors maintain a balance between effective rinsing and water usage.

During pre-rinse, the amount of soils being pushed in all directions by hot water hoses can be overwhelming. If the room air movement is not adequate, within minutes, the sanitor can barely see 6 ft. in front of them. The saying goes: "If you can't see it or reach it, you can't clean it." The pre-rinsing step is a labor-intensive step and, if well executed, will positively influence the outcome of all remaining steps in the process. What does good look like? At the end of pre-rinse, there should be no visible soils on either the equipment or infrastructure. Some individuals may be fooled into believing the room is ready for production at this stage but visible cleanliness at this stage of the process does not necessarily mean that the surfaces and equipment are chemically, physically, and microbiologically clean. A few companies have a supervisor or third-shift QA technician perform a "pre-op" type inspection of the equipment and surrounding area at this point to verify the removal of visible soils before applying detergent. Teamwork is crucial for success. That is, sanitors must be aware of how their actions impact adjacent areas. They cannot work from the center of their area and spray soils indiscriminately and with no regard to adjacent areas outside their immediate area. When sanitors work as a team, everyone sprays soils to common collection zones and never sprays in the direction of cleaned equipment causing recleaning (rework).

(a) Pre-rinse in the direction from top to bottom (i.e., from the top of the equipment down to the floor), including the low ceiling and overhead structures impacted daily by production, and surrounding walls and floors. Use a ladder when required to access the top of equipment. Sanitors must pay as much attention to cleaning the environment surrounding the equipment as they do to the equipment itself. This includes ceiling, walls, floors, drains, and any overhead infrastructure surfaces that can be cleaned, such as pipes and drop cords.

(b) Hot water rinsing must utilize temperatures sufficient to melt the fat soils without baking on proteins (i.e., 54.4°–71.1°/130–160 °F; Cramer 2013), and at a sufficient pressure (low boosted to a maximum of 280 psi at 10 gpm) to remove visible soils, but not at excessive levels that may force soils deeper into equipment or through enclosure seals. The availability of hot water must be sufficient to maintain the target temperature and pressure throughout the facility during peak use. It may be necessary to customize water pressure for specific applications. A multifunctional team composed of individuals of varying expertise and responsibility, such as maintenance, engineering, operations, sanitation, and QA, should review the facility and equipment to determine the optimum cleaning parameters required to maximize cleaning effectiveness without degrading assets.

(c) All hoses used to deliver water, foaming cleaner, and sanitizer must have a proper nozzle and tip. Open-ended hoses will waste water and chemicals. Nozzle tips must be in place to disperse water/chemical appropriately. "Bullet" tips must never be used.

(d) Sanitors should be aware of the direction of spray, taking into account sanitor(s) working in adjacent areas. A well-thought-out plan and training encourage sanitors to work as a team to direct soils to "soil zones" for collection and minimize the need for recleaning areas and equipment due to recontamination from overspray.

(e) Sanitors should avoid using the water hose as a "floor sweeper" and should never use the water hose to "force" soils down the drain or to unblock a plugged drain. Sweep up floor debris periodically using shovels. Never return to cleaning food contact equipment after handling drains without proper hand washing or a change of Personnel Protective Equipment (PPE). In order to reduce the risk of cross-contamination from drains, many companies assign a dedicated individual to clean drains, using a separate, color-coded (generally black) brush and bucket designated and identified for drain use only.

(f) Sanitors must never place parts directly on the floor or on operations floor stands. Dirty and clean parts should not be stored on the same table or cart. Cleaned parts should be positioned on a cart or rack that allows parts to self-drain. Sanitors must never allow sanitation tools that will touch product contact surfaces to rest on the floor when not in use.

(g) All sanitors must self-inspect (with a flashlight) their work when the pre-rinse step is complete. They must inspect from all angles and use all the senses (vision, touch, and smell) to ensure that there are no visible soils present on ANY surface. Focusing on the known "hard to clean" areas will either reveal the need for "spot" rinsing or provide assurance to move on to the next steps in the process. Some companies include detailed photos of equipment in the written Sanitation Standard Operating Procedures (SSOPs) in order to emphasize the cleaning and inspection of these "hard to clean" areas. The target outcome at the end of the pre-rinse step is to have the equipment and area free of visible soil and fat films.

(h) Detergent application must not begin until all sanitors in an area have completed the pre-rinse step and performed a self-inspection.

3. Soap and Scrub

Scrubbing has become a lost "art" and unfortunately, for many companies, this step is only a "soap and rinse." While the chemistry of soap has improved, there is no commercial cleaning product available that has proven claims of "scrubbing bubbles," although many try. Mechanical action is an essential part of the soaping step. While hot water is effective in removing loose soils and melting fats, surface films, including biofilms and allergens, can only be removed through scrubbing. Over time, the lack of scrubbing becomes apparent when stainless steel equipment surfaces "bead" water, appear dull, or have a "blue" hue.

(a) Sanitors should apply the foam in a side-to-side manner to the equipment and surrounding area, including the ceiling, walls, floors, and infrastructure surfaces that can be washed, from the bottom up in the reverse order to pre-rinsing. The consistency of the foam should be adjusted to a shaving or whipped cream consistency so that the soap does not run off a vertical surface when applied, thereby allowing for adequate contact time with these surfaces. Never foam more area than can be scrubbed before the foam dries on the surface (about 15–20 min). Consider 360° access when foaming. This will ensure that ALL surfaces are treated to the ability that the sanitor has access to them.

(b) Ensure that the detergent is compatible with the component composition of the equipment being cleaned and ensure that sanitors are handling chemicals safely and wearing protective equipment such protective eyewear (i.e., goggles or face shields), chemical resistant gloves, sleeves, and/or when handling and applying chemicals.

(c) Scrub all foamed surfaces using proper tools and abrasive action. Applying foam alone will not remove surface organic residues and inorganic stains. Use extension poles and brushes to scrub surfaces beyond your reach.

(d) Always clean drains as part of the "detergent" step and complete the task using dedicated personnel and tools such as a color-coded brush, bucket, and pads.

4. Post-Rinse

Excessive post-rinsing must be avoided. Rinse to remove foam, not debris. Spraying off debris at this stage using pre-rinse water pressure will result in overspray and the transfer of soil particles from one equipment surface to another. It is a true sign that pre-rinse was NOT completed effectively. Switching to the sanitizer hose for "touch ups" will also reduce excessive humidity that results in condensation.

(a) Rinse using the same spraying best practices as described for pre-rinsing above. If possible, reduce water temperature and pressure in order to minimize condensation formation and overspray. Pressure is the easiest to reduce and can be regulated through a valve or reduced through a nozzle tip that disperses the water stream. Many companies have discontinued using any pressurized water for the final rinse, opting instead for a flood rinse, particularly in ready-to-eat or exposed finished product areas. This is due to the increased risk for cross-contamination from non-product surfaces, such as floors and drains, to food contact surfaces through the use of boosted water. Pressurized water causes aerosolization of water-containing soil and bacteria that can then settle on equipment, and facilitates moisture penetration (and contaminants) into sealed areas that can later seep back out onto product contact surfaces and the product.

(b) Self-inspect equipment and surrounding area. Deviations must be addressed by Sanitation BEFORE preoperation inspection.

(c) Sanitation effectiveness testing is performed using a variety of tests including ATP, aerobic plate count (APC), and/or environmental swabbing for target organisms such as *Listeria*. Testing should be performed immediately after the equipment and infrastructure have been rinsed. NEVER "swab" a surface that is NOT visibly clean. ATP swabs are best taken BEFORE sanitizer application. Most companies take all swabs before the application of sanitizer as a measure of the cleaning process and to limit any potential interference of the analytical method by the sanitizer.

(d) Switch air-handling equipment to production mode, if applicable.

5. Pre-Operation (Pre-op) Inspection

   Pre-operation inspection (pre-op) is the verification step to ensure that the equipment and infrastructure are visibly clean. The main objective of pre-op inspection is NOT to find the sanitor's mistakes but rather to confirm that the cleaning phase of the sanitation process was completed in a satisfactory manner so that the transition phase to production can proceed. Sanitors must be accountable for their work through self-inspection and verification by their supervisor prior to pre-op inspection. When fully executed, an effective pre-op inspection is a multilayered audit that provides invaluable information and feedback. Differentiate between a true sanitation "miss" and overspray. Sanitors must be trained to understand where the sanitation misses are, adapt by taking more time and effort to tackle the "hard to clean" areas on equipment, and by prioritizing self-inspection to these areas. Overspray is a consequence of not realizing the impact of the direction of water spray during rinsing. In fact, sanitors must never use a pressurized water hose to remediate findings. Water hoses should be removed from the area after the post-rinse step is completed. Findings can be addressed using mechanical means such as a bucket and paper towel, pad or brush, followed by a low-pressure (flood) sanitizer hose. If a significant sanitation miss is found, and a rewash is needed, this results in a delay in proceeding to the next step in the sanitation process. Everyone performing either self-inspection or pre-op must be encouraged to use all senses including visual, touch, and smell. A flashlight and mirror are critical for visual inspection along with using gloved hands to touch equipment surfaces and being aware of off-odors emanating from equipment or infrastructure.

(a) Pre-op inspection must be conducted by an independent, non-sanitation person and not by the sanitor.
(b) When deviations are found, the sanitor must be present to remediate findings using manual means such as a bucket and a pad or brush and a low-pressure sanitizer spray. Pressurized hot water should not be used unless a complete reclean of the area and equipment is required.
(c) Track and trend deviations. Investigate repeat deviations to determine the root cause. Common root causes include poor equipment sanitary design, poor accessibility to surfaces that need to be cleaned, and subpar performance by inadequately trained and supervised sanitors. Unaddressed repeat findings are an indication of program failure.

6. Condensation Control

(a) Sanitors assigned to tasks involving condensation control must change out of their sanitation apparel and wear Production PPE consistent with the area. Condensation tools dedicated to the task must be handled in a sanitary manner to avoid contaminating overhead surfaces when removing condensation.

7. Sanitization

(a) All equipment and environmental surfaces must be sanitized through the use of a low-pressure, high-volume sanitizer application (flood) at the "no rinse" concentration level. Avoid significant pooling on food contact surfaces. Many companies prefer using a foaming sanitizer to allow employees to easily see where they have and have not applied the sanitizer. The use of foam also allows for the appropriate contact time of the sanitizer on equipment and surfaces (particularly on vertical surfaces such as walls and equipment frames) prior to inspection and equipment setup. Some companies apply the sanitizer in a single step while others use a two-step application method. In the two-step method, which may be used on a daily or weekly basis, or in case of an event or issue, a higher concentration of sanitizer is applied to all equipment and environmental surfaces followed by a low-pressure flood rinse of water to food contact surfaces only, and a reapplication of the sanitizer at the approved "no rinse" level to those same food contact surfaces.
(b) Make sure to separate parts when sanitizing them to ensure complete coverage. Never sanitize parts that are stacked together or piled in a bin or barrel.
(c) The sanitation step should be repeated if significant assembly of equipment has occurred or if a significant amount of time has passed (i.e., maximum of 4 h) prior to production startup.

4.2.2   Non-Daily Sanitation: Deep Cleaning and Intervention

The simplest way to describe the merits of non-daily cleaning is to use the analogy of dental care. In spite of the best efforts to perform dental hygiene daily, trips to the dentist routinely involve the physical removal of plaque and tartar by the dental hygienist. This procedure is preventive in nature since if plaque and tartar are allowed to accumulate, gums will begin to recede and bleed. Untreated, this could result in periodontal disease that can lead to other health complications. In much the same way, daily cleaning of equipment is restricted to surfaces that are accessible. However, equipment can be complex and consist of multiple layers that often overlap each other. Sanitors can only clean those surfaces that they have access to and that time allows. Recognizing this, non-daily tasks are created to further dismantle equipment for deep cleaning. Usually, tasks are localized to multicomponent parts or cabinet enclosures. These tasks are preventive in nature and are scheduled events and not driven by an event or finding. However, an event or finding can trigger an additional or previously unscheduled deep cleaning. Deep cleaning involves the further disassembly of equipment to expose surfaces that are not accessible without the use of tools and maintenance expertise. Deep cleaning tasks are performed after the seven-step sanitation process (typically started after the post-rinse step) and performed without the use of pressurized hot water. Tasks may include, but are not limited to, the following:

- Hand wiping and sanitizing the inner surfaces of enclosures, guards, and covers
- Removal, soaking, and scrubbing of equipment and conveyor belts
- Removal, soaking, and scrubbing ultra-high-molecular-weight polyethylene (UHMW) conveyor guides and supports
- Removal, dismantling, soaking, and scrubbing conveyor drive sprockets, static rollers, conveyor support structures, and equipment components that are composed of multiple pieces that are overlapped or bolted or pressed together (sandwiched)

An intervention for an event or finding often involves a non-daily task and treatment of the entire piece of equipment as a whole. The two most common forms are a total teardown followed by intensified deep cleaning of the equipment and/or the use of a heat intervention such as steam treating or baking equipment. The key objective is to expose and treat ALL surfaces. In a teardown procedure, all mated or sandwiched points on equipment are opened up to expose surfaces and allow them to be physically cleaned and sanitized. In using heat, a conduction is used to heat all surfaces to temperatures that are lethal to pathogens and spoilage organism. A temperature of 160 °F (71 °C) with a hold time of 20–30 min is recommended to treat equipment for pathogens such as *Listeria* (Tompkin 2002). However, other microbes such a lactics may require higher temperatures or longer hold times, as some of these organisms tend to be more heat-tolerant. Validation and verification of these processes will help establish the most effective parameters.

As noted at the annual North American Meat Institute (NAMI) Advanced *Listeria* course, the use of heat intervention for equipment is still not a common practice amongst participants and its use is sometimes a subject of debate in the industry. However, for those companies that have successfully implemented heat intervention, there are a several key points to note:

1. The process used must be thoroughly assessed by a multifunction team prior to implementation. There are human safety and equipment safety concerns that must be considered, particularly when using saturated steam. The process must be both effective and protect assets, such as electrical components and non-metal pieces from damage.
2. Temperature probe placement is key. Probes must be placed on surfaces that represent the cold (hardest to heat up) and hot (easiest to heat up) spots. Do not hang probes in the air, as air temperature does not accurately represent equipment surface temperature. The duration of time for the procedure to be effective must be validated.
3. Only equipment components that can withstand the heat treatment are included in the physical "tent." Specifically, if ultra-high-molecular-weight (UHMW) pieces can be removed, then do so since the softening point of this material will be reached during steaming. If there is a chance that steam will penetrate adjacent cabinets, block access, and keep adjacent cabinets under positive pressure using compressed air.
4. Introduce steam through a manifold that provides even dispersion throughout the tented area.
5. An efficient steaming process will quickly even out the cold and hot spot temperatures and there will be no evidence of steam leakage.
6. The balance between efficacy and effectiveness is based on a right combination of time and temperature.
7. Moist heat is far more efficient than dry heat.
8. Tenting is a great option when large equipment is stationary. For equipment that can be moved, such as slicers, tables, etc., an oven (e.g., smokehouse) and heating with moist heat is another good option. Some companies have a routine rotation and an assigned and validated oven schedule for heating "problem" pieces of equipment.
9. For small parts, a hot water COP tank or boil tank is a good option for heat treatment.

All equipment and surrounding infrastructure must have a non-daily sanitation task that can be accomplished during non-production time that allows either further disassembly or accessibility of equipment requiring maintenance support. This may require operations to temporarily move stored product or equipment from areas to be treated. These tasks are preventive in nature and are captured as part of an MSS that schedules non-daily sanitation tasks for the entire year. Weekly planning and schedule compliance are discussed as an agenda item at the regular sanitation meeting. Ideally, these tasks are coordinated with routine preventative maintenance tasks.

4.2.3   Operational Sanitation

Good manufacturing practices (GMP) are well established in the food industry. Operational sanitation tasks are a specific subset of sanitation best practices that ensure that exposed products are handled and processed on equipment and in surrounding areas that are clean and sanitary throughout the entire production "window." This becomes a challenge as the production environment is in a constant state of movement of product, supplies, equipment, and people. Products are most vulnerable during exposure to the environment, prior to primary packaging. In the case of ready-to-eat (RTE) cooked products, the risk extends from the point immediately after thermal lethality step until the time the product is packaged. For some products such as fresh produce, this point may begin following the application of an antimicrobial spray or flume. Procedures must be in place to mitigate the risk of contamination associated with the following:

- Direct and indirect contamination by front-line employees
- Product accumulation in equipment leading to product contamination
- Condensation leading to food contact surface or product contamination

The best practice for operational cleaning of equipment within the production window is to replace "dirty" equipment parts with "clean" ones, if available, and to clean dirty parts in a wash area segregated from the production area. Plants should never use pressurized hot water to clean within the production area, especially when production is running in the area, such as on an adjacent line or piece of equipment. The risk associated with overspray to adjacent equipment, from cross-contamination from floors and drains, and the risk of forcing soils deeper into equipment is too high. The use of hot water will also increase the risk of room condensation. Dry cleaning with compressed air is also not recommended. If a mid-shift cleanup cannot be avoided, the best option would be dry cleaning followed by the use of a low-pressure/high-volume sanitizer at the "no rinse" concentration.

The routine use of a hand sanitizer to sanitize front-line employees' hands and an approved sanitizer to sanitize food contact surfaces are also good options for maintaining the sanitary condition of product handlers and surfaces. Several companies have made this practice routine by activating an audible buzzer, or flashing light, every 15 min to remind employees to sanitize their hands, tools, and work station. Alternately, some companies assign an employee to walk around and apply sanitizer to employees' hands and to surfaces. Sanitizers require sufficient contact time to be effective. "Spray, wait, and wipe" or "spray, wipe, and spray again" should be practiced without allowing the sanitizer to pool on food contact surfaces.

4.3   Sanitary Design and Access

While no piece of equipment is perfect, equipment engineers and manufacturers have improved equipment design in recent years. There are many standards to refer to for equipment and facility design including the 3-A sanitary designs used for dairy equipment, the European Hygiene Engineering and Design group (EHEDG) guidelines for equipment and infrastructure, and the NAMI sanitary design principles and checklists for equipment and facility design (NAMI 2016). The NAMI design principles offer a practical, simple, and effective approach and include easy-to-use assessment tools and checklists. These checklists were originally designed to address pathogenic microorganisms; however, the same rules and best practices are universal and effective when addressing shelf life concerns as well. One point to keep in mind is that the "score" is irrelevant. It is much more important is to determine the marginal and unacceptable attributes of the equipment and decide during fabrication what issues can be addressed through a redesign. If the sanitary design issue cannot be "designed out," the plant must develop procedures to mitigate the risk associated with the condition such as more frequent disassembly and deep cleaning. Progressive companies will bring the supplier to the facility after installation and following several cycles of production/sanitation, to determine whether the equipment is cleanable in the long term and under actual conditions of use.

One consideration that is often overlooked is equipment access for sanitation. Equipment access for operators comes in the form of stairs and platforms with the assumption that this level of access meets the needs of sanitation. In most cases, the level of access required for sanitation is significantly higher. Operating a piece of equipment is often performed in a fixed operator/equipment interface; however, the sanitor requires access from all sides of the equipment. For example, if sanitors only have access to the operator side of the equipment, how does the sanitor clean the opposite side? Sanitors must be able to clean from the top of the equipment down to the floor as well as the undersides of equipment. If adequate access to overhead and elevated areas is not provided with sufficient height, sanitors must use ladders or, worse, stand on handrails or not clean from an elevated position. Equipment clearance from the floor must be sufficient to allow sanitors to access the undersides and surrounding floor. This level of access is also important from an inspection standpoint.

Clean In Place (CIP) systems in equipment such as ovens and freezers should never rely only on automated systems for cleaning. The individual responsible to operate the CIP must be trained to verify the CIP cleaning task and remediate manually as required.

4.4 Verification and Performance Swabbing

First, let us agree that "clean is clean." Regardless of type of facility or area within the facility, the definition of "clean" must be universal. That applies to all aspects of performance measurements. What can vary are the conditions that ultimately trigger a corrective action. What is most important is to not allow repeat findings with no corrective actions. There is no better way to demonstrate the ineffectiveness of the plant's program to auditors/regulators and to send a message to the sanitation team that nobody cares than a failure to implement effective corrective actions. A "Seek and Destroy" approach (Malley et al. 2015) used for addressing *Listeria* findings can also be applied for determining the root cause for shelf life issues and implementing corrective actions that are both effective and sustainable. This may include but is not limited to an evaluation of sanitary design issues within the facility or equipment, poor equipment access, and/or sanitation team performance. Performance is often regarded as the last option however, depending on the level of training, and results of key performance indicators, performance should not be ruled out completely.

There are three types of performance measurements with each one serving a specific role:

1. Visual Inspection (Pre-op): The pre-op inspection is performed by someone who is independent from sanitation, such as a QA technician. Each piece of equipment and the surrounding area and infrastructure is inspected after post-rinse for the presence of visible debris missed by sanitation or for overspray debris. Successful completion of a pre-op inspection allows the completion of tasks required to transition the room from sanitation to production. Pre-op performance must be a measurement of what the sanitor has failed to effectively clean and what the inspector has failed to find.

2. Residue Monitoring (ATP bioluminescence): In this case, a sanitation supervisor or designate swabs a selected number of equipment sites (usually zone 1) after post-rinsing but prior to sanitization. The results are obtained within minutes and measure the presence of residue that includes both organic residue (food soils) and in some cases, microbes. Frankly, there are mixed views on the merit of this testing. ATP is generally viewed as a qualitative tool used for training since there are no data to demonstrate that results are comparable to microbial testing, particularly at lower levels. To be quantitative, the tool must be able to perform well in a measurement system analysis (MSA) that measures both repeatability and reproducibility. This means that the same sample site will obtain consistent results tested a number of times and that multiple units testing the same sample site will yield comparable results.

3. Microbial Surveillance (APC, Lactics, *Listeria* spp., etc.): The type of microbiological testing used to monitor and verify the sanitary conditions of the equipment and environment and the tests performed may vary by product, process, and objective for testing. Those tests often used as indicators of sanitary conditions that may affect shelf life include but are not limited to aerobic plate counts (APC), lactics, or Enterobacteriaceae (EB). The QA technician or designate swabs a selected number of equipment and area sites (all zones) immediately after the equipment and area have completed the final sanitization step. Depending on the method used for analysis and target organism, microbial testing is often viewed as a lagging indicator, or historical data, since the results may be received anywhere from 24 to 48 h or as much as a week from the time the samples are taken.

Post-rinse sampling is an investigative approach to finding potential areas of equipment or infrastructure harborage within the sanitation process. While APC swabs can be used during this investigation, many companies prefer swabbing specifically for organisms such as *Listeria* for food safety issues or for lactics when spoilage is a concern. Swabs sites are typically on equipment support legs near the floor juncture or the floor surface underneath equipment either after the pre-rinse or post-rinse. The process of rinsing enables water to penetrate equipment surfaces far beyond what is dismantled or accessible on a daily basis. Close proximity to floor drains in freezers/ovens or persistent wet spots along floor wall junctures are other good examples of sample sites. While a positive result will not indicate precisely where the point of harborage is, it will notify the team that there is an active shedding of a harborage site that requires immediate attention. The advantage of this approach to sampling is that while sanitation sampling at time zero or : Environmental Monitoring Program (EMP) sampling during production are transfer point detection (you found something on a surface but it is most likely not where the contamination originated from), post-rinse sampling is harborage site detection (a growth niche is occupied and is shedding). Post-rinse sampling detects issues at the area where "shedding" occurs and these sites are distinctly different from typical EMP and sanitation effectiveness sites since harborage sites are not readily available for sampling.

4.5 Maturity Model

Many companies use maturity models to help them see how their programs compare in maturity to industry best practices. In this case, the model outlines the progress and current state of sanitation in the food industry. A manufacturer can use the model to see where their sanitation programs currently stand in relationship to top performers in the industry and shows them how they can move through the different levels of maturity toward continuous improvement and optimization of sanitation, leading to improvements in product food safety and quality (Table 4.1).

Table 4.1 Sanitation maturity model

Low state	Medium state	High state
Sanitation process—Sanitation is an organized process that is respected and supported		
1. Sanitation documentation meets regulatory and audit expectations but sanitors are not properly trained or held accountable for their work. Supervisors cannot oversee their team since they are often working themselves	1. SSOPs provide detailed cleaning instructions and describe performance expectations. Sanitors are actively supervised for their individual workmanship and teamwork	1. SSOPs are routinely verified, updated, and audit ready. Sanitor and supervisor training is ongoing and documented through training records. Sanitors use flashlights to self-inspect their work after each rinse step
2. Sanitors work as individuals, do not stay in sequence, and are not aware or care about the consequence of overspray	2. Sanitors work together to stay in sequence and minimize the recontamination of adjacent equipment or area caused by overspray	2. Sanitors work in a predetermined direction that push soils to collection areas for ongoing pickup instead of directing soils to drains

(continued)

Table 4.1 (continued)

Low state	Medium state	High state
3. Rooms are expected to be released back to production on time regardless of the delays in starting sanitation caused by extended production or room transfer tasks that are poorly completed by production	3. Operations, maintenance, and QA support sanitation to transition areas from production to sanitation and vice versa	3. Cross-functional room transfer tasks are verified by signed check sheets. The duration allotted to sanitation is consistent regardless of sanitation startup delays

Non-daily tasks—Tasks are deployed beyond daily procedures to address the risk associated with soil type and equipment complexity

Low state	Medium state	High state
1. There is no formal requirement for non-daily tasks other than to coordinate equipment cleaning with preventative maintenance tasks as specified by the Original Equipment Manufacturer (OEM)	1. Every piece of RTE equipment has a non-daily task that is scheduled based on the prevention of findings. There is an awareness of the benefits of heat treatment but this practice has not been fully deployed	1. RAW and RTE equipment have non-daily tasks. Heat intervention or full equipment teardown is fully deployed. Plants modify existing equipment based on findings and design new equipment to eliminate known growth niches and to simplify non-daily tasks
2. If equipment requires intervention due to issues related to food safety or quality, the SSOP is altered by increasing the frequency of the procedure, using higher chemical concentrations or introducing new chemicals	2. Deep cleaning and teardown tasks are fully developed. The schedule of non-daily tasks is captured in the MSS	2. Non-daily tasks are effective without degrading assets. The frequency of deep cleaning and intervention tasks is optimized by using data to determine when tasks are required
3. The MSS is not overtly visible and does not have expectations or oversight	3. Non-daily task planning is reviewed at the weekly sanitation meeting. Production requirements may postpone tasks requiring them to be rescheduled	3. Non-daily task completion and planning is reviewed at the weekly sanitation meeting. Task completion is a Key Performance Indicator (KPI) and if tasks are delayed, there is an escalation process for resolution

Sanitary design—Sanitary design and access are built into equipment design and installation

Low state	Medium state	High state
1. Design is a consideration for the procurement of equipment	1. Sanitary design is a factor in the procurement of equipment by engineering and the cross-functional plant team	1. Sanitary design is embedded as a requirement for the procurement of equipment. Suppliers and contractors receive preferred status based on their ability to execute against these principles

(continued)

Table 4.1 (continued)

Low state	Medium state	High state
2. Engineering relies on the OEM to provide equipment that meets sanitary design principles. Tools such as the NAMI sanitary design checklist are used to generate a compliance score. Equipment installed is "off the shelf"	2. Tools such as the NAMI sanitary design checklist are used by key internal stakeholders to determine the marginal and unacceptable conditions of the equipment design. There is a bias toward managing deficiencies through procedures rather than redesign deficiencies that would add cost or delay delivery	2. Sanitary design is reviewed by key internal stakeholders at every key stage between procurement and installation. Deficiencies are designed out where possible
3. Access to all equipment surfaces and access to internal structures for sanitation is not considered prior to installation	3. Equipment access for sanitation is a consideration. Issues arising after installation are managed. The use of ladders and tools to provide access during sanitation is an acceptable outcome	3. Platforms and access to clean equipment from all sides is designed in to avoid the need for ladders during sanitation. Guard removal to access internal surfaces do not require tools
Performance—Driving continuous improvement		
1. Sanitation costs are the only performance metric that has routine tracking and oversight	1. Performance metrics beyond costs are tracked, trended, and reviewed at the scheduled sanitation meeting. Issue resolution drives continuous improvement	1. Sanitation KPIs are regularly reviewed at the executive level. Independent verification and technical support drive program robustness and value
2. Swabbing is biased toward achieving results that meet corporate targets	2. Swabbing is biased to finding problem areas and verifying that corrective actions implemented are both effective and sustainable	2. Swabbing is biased to verifying control of intervention "hurdles." Swab data are valued and used as a predictive tool to ensure that control measures are working
3. Pre-op inspection is used to point out what the sanitor has missed. When deviations are found, they are addressed using the pressurized hot water hose	3. Pre-op inspection is a coaching moment between the sanitor and the auditor. Findings are addressed by manual means or through the use of the sanitizer hose	3. Pre-op is a multilayered verification process where findings (sanitor) and missed findings (auditor) are tracked. Repeat findings are addressed through changes in equipment design/access or through performance accountability

4.6 Summary

Sanitation is a key enabler for manufacturing plants to produce safe products that will consistently meet shelf life. It is a defined process that requires the same level of supervision and support that production processes receive. Sanitation times must be respected and the non-daily tasks must be deployed at a frequency that is appropriate, given the complexity of equipment, the type of soils present, and without degrading the assets affected. Sanitation metrics must go beyond costs to measure effectiveness and efficiency of cleaning. Equipment design must permit cleaning and provide access so that sanitors can do their work safely and effectively. Finally, satisfactory performance must be driven through personal accountability and verified following a multilayered audit approach. Repeat deviations in effectiveness metrics must drive root cause analysis, leading to effective and sustainable corrective actions.

One last point is that there is a talent gap today in the sanitation industry. The lack of talented floor supervisors and sanitation managers is an indication that as leaders, we have failed to invest the effort to attract and retain sanitation talent. While many plants have chosen instead to use third-party sanitation contractors, plants must retain control over and manage the sanitation program, holding those accountable to perform sanitation tasks regardless of whether the labor is internal or contracted. Food plants that recognize these human resource challenges, and devise strategies to recruit, train, and retain sanitation personnel are more likely to succeed in operating an effective sanitation program.

References

Allen, J.R., and E.M. Foster. 1960. Spoilage of vacuum packed sliced processed meats during refrigerated storage. *Journal of Food Science* 25: 19–25.

Cramer, Michael M. 2013. *Food Plant Sanitation second edition*. Boca Raton, FL: Taylor & Francis.

Doyle, Marjorie Ellin, and Kathleen E. Glass. 2010. Sodium reduction and it's effect on food safety, food quality and human health. *Comprehensive Reviews in Food Science and Food Safety* 9: 44–55.

Kempton, A.G., and S.R. Bobier. 1970. Bacterial growth in refrigerated, vacuum-packaged luncheon meats. *Canadian Journal of Microbiology* 16: 287–297.

Malley, T.J.V., J. Butts, and M. Wiedmann. 2015. Seek and destroy process: *Listeria monocytogenes* Process controls in the ready-to-eat and poultry industry. *Journal of Food Protection* 78: 436–445.

North American Meat Institute (NAMI). 2016. *Sanitary equipment design principles: checklist & glossary. Sanitary Equipment Design Taskforce*. Washington, DC: AMI Foundation. https:// www.meatinstitute.org/ht/a/GetDocumentAction/i/97261. Accessed 16 February 2018.

Taormina, Peter J. 2010. Implications of salt and sodium seduction on microbial food safety. *Critical Reviews in Food Science and Nutrition* 50: 209–227.

Tompkin, R.B. 2002. Control of *Listeria monocytogenes* in the food-processing environment. *Journal of Food Protection* 65: 709–725.

Chapter 5
Advanced Processing Techniques for Extending the Shelf Life of Foods

Sarah M. Hertrich and Brendan A. Niemira

5.1 Introduction: What Is Food Spoilage and How Is it Prevented?

Spoilage is the development of unwanted characteristics of color, texture, odors, flavors, or other organoleptic or nutritive qualities that make food less desirable (Montville et al. 2012). Spoilage can be the loss of attractive attributes or the development of objectionable attributes. These characteristics can develop through abiotic chemical or physical changes in the food (e.g., oxidation and drying), through metabolic activity of the commodity itself (e.g., over-ripening and respiration), or through microbial activity from native or contaminating bacteria, fungi, or viruses. There are many stages within the food production process in which spoilage can occur, including processing, packaging, distribution, retail display, transport, storage, and during handling by the consumer (Gould 1996). One of the many goals of the food industry is to delay the onset of food spoilage by extending the amount of time food can remain in storage before it will become spoiled. Current research continues across the globe to develop cost-effective food processing techniques that will help create food products of high quality that are safe and readily storable.

This chapter will focus primarily on food preservation techniques that prevent or inhibit microbial growth, and the unwanted resulting changes in the food product. One of the most familiar types of preservation is refrigeration. Refrigeration allows for low-temperature storage of food products including meat and dairy, which keeps present microorganisms "at bay" and preventing current populations from multiplying. Other bacterial inhibition techniques include freezing, drying, curing, vacuum packaging, modified atmosphere packaging, acidifying, fermenting, and addition of

S. M. Hertrich · B. A. Niemira (✉)
Food Safety and Intervention Technologies Research Unit, Eastern Regional Research Center, USDA-ARS, Wyndmoor, PA, USA
e-mail: shertrich@naccme.com; Brendan.Niemira@ARS.USDA.GOV

© Springer Nature Switzerland AG 2021
P. J. Taormina, M. D. Hardin (eds.), *Food Safety and Quality-Based Shelf Life of Perishable Foods*, Food Microbiology and Food Safety,
https://doi.org/10.1007/978-3-030-54375-4_5

preservatives (Gould 1996). Other preservation techniques aim at inactivation or elimination of microorganisms in food products such as pasteurization, sterilization, and irradiation (Gould 1996). There are also techniques that prevent the entry of microorganisms into foods such as aseptic packaging techniques (Gould 1996). Consumer demands for less heavily preserved, unprocessed foods of high quality have led to the need for processing technologies that will (1) provide less damage to the product; (2) preserve the organoleptic and nutritional value of the produce without the use of heat (i.e., nonthermal processing); (3) allow the use of more natural antimicrobials (i.e., essential oils and spices); and (4) provide different packaging methods that will extend shelf life (i.e., modified atmosphere packaging) (Gould 1996). This chapter will examine novel food processing technologies, and place them in context with conventional techniques used to preserve a variety of commodities. The emphasis will be on spoilage and food quality, with additional discussion of relevant food safety applications.

5.2 Processing Techniques for Extending Shelf Life

5.2.1 Basic Preservation Techniques

5.2.1.1 Cold Storage/Cooling/Freezing

Refrigeration, also known as cool or cold storage/chilling, refers to the storage of foods at temperatures from 16 to −2 °C (Montville et al. 2012). The mechanism by which refrigeration is able to delay food spoilage is by decreasing chemical reaction rates and subsequently delaying most microbial growth (Montville et al. 2012). The shelf life of meat, fish, and poultry products can be extended from 1 day to up to 2 weeks with refrigeration (Farkas 2001). The shelf life of fruits and vegetables can be extended from 1–50 days to up to 300 days with refrigeration, depending on the product (Farkas 2001).

It is generally recommended that foods are kept out of the "danger zone"—the temperature range which favors microbial growth in foods which is between 40 and 140 °F (USDA 2013). To keep food safe from the growth of pathogens or spoilage microbes, hot foods should be kept hot (at or above 140 °F) and cold foods should be kept cold (below 40 °F). If a food reaches temperatures in the danger zone, bacterial populations could double in only 20 min. At colder temperatures, some concerns may remain regarding psychrotrophic bacteria, also known as "cold-tolerant" bacteria, which include *Yersinia, Vibrio, Pseudomonas, Listeria,* and *Aeromonas* species. These types of organisms are made up of more unsaturated fatty acid residues and more branched-chain fatty acids in their lipid molecules compared to their mesophilic counterparts, allowing their membrane proteins to be able to function at lower temperatures (Montville et al. 2012). While these species are able to survive at lower temperatures, growth is typically slowed which helps to extend the shelf life. The ability of psychrotrophic bacteria to alter their fatty acid composition to maintain membrane fluidity is known as homeoviscous adaptation (Montville et al. 2012).

Another well-known basic form of food preservation is freezing. In general, foods are initially frozen by contact with cold air, contact with a cold surface, or submersion into a refrigerant liquid such as liquid nitrogen then stored at –20 °C (Montville et al. 2012). Depending on the amount of water that is present within a food, the eutectic temperature (the temperature where the solutes within the food reach their solubility limit and the available water freezes) will vary. This is the state which the food reaches the totally frozen state and occurs at –15 to –20 °C for fruits and vegetables and –40 °C for meats (Montville et al. 2012). Some components of foods, such as sugars and transmembrane proteins, can protect cells from mechanical injury during freezing. During freezing, many microbes go into osmotic shock where ice formation within the cells of the organisms causes mechanical injury, resulting in reduction of populations. However, some pathogens are able to withstand extended frozen storage without suffering complete kill (Niemira et al. 2002; Niemira et al. 2003). Microbial growth can occur during thawing when the contents of freeze-ruptured food tissues become available for metabolism by associated microorganisms. In a situation of cycles of freeze–thaw, microbial populations take advantage of the available nutrients and present a greater risk. Refrozen products that have been thawed are therefore especially susceptible to microbial spoilage. Freezing can also alter the sensory qualities of the food. Development of ice crystals on the surface of foods can occur during temperature fluctuation, or freeze–thawing, which can also lead to microbial spoilage.

5.2.2 Thermal/Heating

Thermal processing for shelf life extension should be regarded as distinct from cooking. Cooking transforms raw ingredients, enhancing digestibility, improving flavor, and eliminating potentially harmful contaminating organism. However, once cooked, either in the home or at a commercial at point of service, the issue of shelf life is not of primary concern, as the intention is for near-immediate consumption. Thermal processing which is intended to extend the shelf life of food will be combined with specialized packaging. Historically, these have been metal cans or glass bottles; more modern packaging for thermally processed foods include formed plastic bottles or jugs, polylaminate plastic bowls, bags, or pouches, or compound containers incorporating foils, laminated paper, or other advanced materials. These containers can readily achieve the familiar shelf-stable "canned" state, or, as with more sensitive commodities and lower thermal processing regimes, be intended for subsequent refrigerated storage.

Heating is the most widely utilized method for killing microbes in food. Louis Pasteur, also referred to as the first food microbiologist, was the first to show that spoilage of milk, wine, and beer could be prevented by heating for a short time at a relatively low temperature. This process is now known as pasteurization and is defined as a mild heat treatment that aims to kill nonspore-forming pathogenic bacteria as well as inactivate enzymes and spoilage bacteria. For example, milk is pasteurized at 161 °F (72 °C) for 15 s. Exposing the milk to high temperatures for longer periods of time will lead to the development of off-flavors. This is true for most foods as many of us have experienced; when a food is heated for too long (or burned), it becomes unpalatable. This is why temperature and time are important variables to consider when utilizing thermal processing.

Another common form of thermal processing is sterilization which aims to kill all microbes present within a food product. This type of thermal processing is typically used in products where the presence of bacterial sporeformers is of concern. Sporeforming bacteria, including some *Clostridium* and *Bacillus* species, can cause severe disease in humans and are highly resistant to heat when in the dormant sporulated state (Setlow 2003). Sporeforming pathogens are only a threat when they germinate and are able to produce endotoxin (Markland et al. 2013a). The goal of commercial sterilization in food products is to eliminate any vegetative or germinating bacteria present and to prevent present spores from germinating in a food product as well as extend the product's shelf life by eliminating spoilage bacteria. Note that commercial sterility is not the same as absolute sterility, which is a process aimed at eliminating all microorganisms within a product, not just pathogens.

5.3 Advanced Processing Technologies

5.3.1 High Hydrostatic Pressure

The use of pressure in food processing was first investigated in 1899 by Bert Hite at the Agricultural Experimental Station of West Virginia University where he discovered that pressure treatment of milk and fruit-based foods could increase their shelf life (San Martin et al. 2002). HHP-processed foods were first introduced to the Japanese market in 1990 and have slowly since been introduced in other countries (San Martin et al. 2002). The technology works by placing food inside of a flexible sealed pouch that is then placed inside of a sealed vessel filled with water that is subjected to high pressure (Montville et al. 2012). The water inside of the vessel acts as the pressure-inducing medium. Not all food products are good candidates for HHP due to the macromolecular changes that may occur during HHP. For example, a food product that is porous or contains internal voids, such as peppers, melons, or raspberries, will become compressed, altering the mouthfeel and sensory qualities of the product. Foods with a higher moisture content tend to be better candidates for HHP because they can withstand the compression (Montville et al. 2012). Currently, HHP is used to commercially process products such as guacamole, pre-sliced deli meats, juices, and oysters. Researchers and producers are interested in expanding the application of HHP to a wider variety of foods in order to increase the safety and shelf life of these foods (Markland 2011; Lou et al. 2012).

According to the research that has been done over the years, microorganisms are inactivated by HHP through a variety of mechanisms. One hypothesis is that a pressure-induced decrease in cell volume can lead to cell leakage and death (San

Martin et al. 2002; Chilton et al. 1996); however, the mechanism of inactivation of cells by HHP can be dependent on the type of pressure applied (cyclic or continuous), temperature, treatment time, strain, cell shape, Gram stain type, growth stage, and treatment medium (San Martin et al. 2002). HHP of vegetative bacterial cells is generally more effective at higher temperatures unless the bacterial species contains the heat shock protein (Hsp), in which case heat would cause a baroresistant effect (Iwahashi et al. 1996; Li and Gänzle 2016). The resistance of microorganisms to HHP is largely dependent upon the species and strain of the microorganism and is extremely variable (Raso et al. 1998), but most vegetative cells of bacteria and yeast are generally inactivated at pressures around 300–400 MPa at ambient temperature (Knorr 1995).

One of the current disadvantages of HHP is its inability to inactivate bacterial spores by pressure alone without altering the sensory qualities of the product (San Martin et al. 2002; Black et al. 2007). Studies show that HHP in combination with heat can inactivate spores within a food product; however, the nutritional and organoleptic qualities of some foods cannot withstand the thermal treatment (Paidhungat et al. 2002; Wuytack et al. 2000). Complete inactivation of spores remains a top priority for high-pressure food processors. It is important to understand the physiology of spores, especially those that pertain to spore inactivation by HHP (Black et al. 2007).

5.3.2   Ultraviolet and Pulsed Light

Application of ultraviolet light (UV) was first used in France for disinfection of drinking water (Masschelein 2002). UV is currently used for the preservation of solid and liquid foods. Because of its limited wave penetration, it is most effective for the inactivation of microorganisms on the surfaces of foods or clear liquids. It has been often used as an alternative to pasteurization of liquids and juices because the lack of heat helps to preserve the fresh flavor as well as extend the shelf life. Pulsed light is another method of food preservation that uses intense and short pulses of broad-spectrum white light. One of the advantages of pulsed light over UV light is that it has a greater penetration depth (Bialka et al. 2008).

UV uses an electromagnetic spectrum from 100 to 400 nm and can be classified into four spectrum regions including UV-A (315–400 nm), UV-B (280–315 nm), UV-C (200–280 nm), and vacuum UV (100–200 nm) (Keklik et al. 2012). The UV-C class is considered to be the most effective regarding microbial contamination of food products (Keklik et al. 2012). The wavelength of pulsed white light ranges from ultraviolet to infrared (Li and Farid 2016). The mechanism by which UV and pulsed light inactivate microorganisms is by damaging the organism's DNA (Li and Farid 2016).

Advantages of using UV and pulsed light technology for food preservation include the relatively low cost and the lack of the generation of chemical residues within the product (Hijnen et al. 2006). Like other nonthermal technologies, one of the limitations of UV light technology is that it has little effect on the killing of bacterial spores; however, it has been shown to help improve the lethal effects of other subsequent treatments (Li and Farid 2016; Gayán et al. 2013). Pulsed light, on the other hand, has been found to effectively inactivate spores in some products (Li and Farid 2016). This is most likely due to the thermal stress that is induced upon the spore coat leading to structural damage of the cell (Hijnen et al. 2006). Another one of the disadvantages of UV light technology is the limited amount of transmittance through products (Gomez-Lopez et al. 2012; Keklik et al. 2012). To help overcome this, UV food treatment chambers have been designed so that fluids flow through it in a thin layer (Gomez-Lopez et al. 2012). Some of the major applications of UV in the food industry include treatment of liquid foods specifically juices, milk, and liquid egg products (Li and Farid 2016). Pulsed light is less widely utilized in the industry, although promising results have been observed for inactivation of bacterial spores in cornmeal (McDonald et al. 2000) and sucrose syrup (Chaine et al. 2012). More studies need to be performed for the optimization of pulsed light technology for the preservation of foods.

5.3.3 Ultrasonication

Ultrasonication is a more recently developed technology that preserves foods through a phenomenon known as cavitation that is created by ultrasound waves (Delmas and Barthe 2015). Cavitation occurs when vapor cavities, or bubbles, are formed within a liquid and the ultrasonic energy continues to increase until the vapor cavities begin to rapidly collapse (implode), creating a shock wave that leads to short periods of high temperature and pressure throughout the liquid (Butz and Tauscher 2002; Montville et al. 2012; Delmas and Barthe 2015). The tiny bubbles which are filled with heat and reactive oxygen species including hydrogen peroxide are released through the breakage of chemical bonds and are able to kill microbial cells (Montville et al. 2012). There are two types of ultrasound including high frequency (megahertz range) and low frequency (kilohertz range) (Montville et al. 2012). The frequency necessary to inactivate microorganisms through ultrasonication is 20–100 kHz (Chandrapala et al. 2012).

While this technology was first described to have the potential to kill microorganisms in 1933 (Szent-Gyorgyi 1933), its potential as a technology to preserve foods was not recognized until the last few decades (Li and Farid 2016). This technology is not effective for the inactivation of microorganisms when used alone; however, it can be effective when used synergistically with other technologies, specifically thermal technologies (Li and Farid 2016). It has also been shown to increase the efficacy of HHP (Raso et al. 1998). The use of ultrasonication along with thermal treatments (thermosonication) has been shown to help reduce the time needed to sterilize a product, which helps to maintain the nutritional and organoleptic qualities of the product (Chandrapala et al. 2012; Mason et al. 2015). Thermosonication studies demonstrated that ultrasonication combined with thermal treatment at 70 °C significantly enhanced the inactivation of *Bacillus cereus* spores in skim milk, liquid beef, cheese products, and rice porridge compared to heating the products alone (Evelyn and Silva 2015).

Because the effects of cavitation are stronger than the adhesion forces (van der Waals attraction) on surfaces, ultrasonication may be most effective at surface disinfection of foods specifically fresh produce or processing equipment (Montville et al. 2012). It could potentially also serve as a good preservation process for liquid products including liquid egg products and juices. The use of the technology is currently expanding in the wine and beverage industries as well (Montville et al. 2012). While ultrasonication shows promise for use as a preservation technology in foods, more studies are needed to address its limitations.

5.3.4 Irradiation

Irradiation uses electrons or photons of sufficient energy that ionizes the molecules they contact. This ionization is usually enacted in water molecules which are the single greatest component of foods, leading to the creation of reactive hydrogen and hydroxyl radicals (Niemira 2014). Unlike UV, which is not powerful enough to ionize and has limited penetrability, ionizing radiation can penetrate solid foods to a depth of 4–6 cm in the case of electrons, but penetrating as much as 40–50 cm in the case of photons (either gamma rays or X-rays). The three types of ionizing radiation are generated by different methods. Gamma rays are produced by certain radioisotopes, with cesium-137 and cobalt-60 being the most commonly used for food processing. Ionizing electrons (known as electron beams or "e-beams") are produced by using magnetic fields to accelerate electrons to high energies, up to 10 MeV. X-rays are created by directing accelerated electrons into a metal target, usually a dense alloy of tungsten. The interaction of the electrons with the metal releases a shower of high-energy X-rays.

All three types of irradiation have been shown to extend the shelf life and improve the safety of fruits, vegetables, meats, poultry, seafood, eggs, and processed foods (Niemira et al. 2002; Alvarez et al. 2006; Niemira and Cooke 2010). Irradiation may be used at doses up to 1 kGy to delay sprouting, slow ripening and/or maturation, and extend shelf life of stored produce. Other commodities have different statutory dose limitations, depending on the purpose of irradiation. For example, iceberg lettuce and spinach can be treated with up to 4.0 kGy to improve safety (CFR 2017).

5.3.5 Cold Plasma

An emerging technology for food processing is cold plasma, which for practical purposes may be regarded as a form of ionized gas. Cold plasma is generated by ionizing gases or gas mixtures with high-voltage electricity or microwaves; the resulting plasma serves as a nonthermal antimicrobial intervention for foods (Niemira et al. 2014). The antimicrobial modes of action are primarily related to (1) chemical reactions with cellular structures; (2) UV damage of cellular components; and (3) UV-mediated DNA strand breakage. A wide array of equipment designs to generate cold plasma have been reported, and are an area of active research and development. In some cases, this research seeks to adapt mature thermal plasma systems used for surface treatment applications in industries such as textiles, printing, polymer processing, or electronics (McHugh and Niemira 2016). Cold plasma-mediated inactivation of human pathogens is a primary goal of these food safety research efforts, with associated extension of shelf life as a concomitant goal (Lacombe et al. 2015; Min et al. 2016).

Key areas for further development of cold plasma include the determination of the precise modes of action for various operating conditions; optimization of feed gas mixtures mated with optimized equipment designs for antimicrobial efficacy, cost and efficiency; and improving compatibility with existing food handling and packaging systems. Another body of emerging body of work is the use of cold plasma to sanitize packaging materials, food contact surfaces, and areas where sanitizer-resistant pathogen biofilms form (Niemira et al. 2014).

5.3.6 Pulsed Electric Fields

The first interest in the use of electric fields for food processing began in the 1930s, before the advent of energy-efficient electronic control systems (Montville et al. 2012). Modern pulsed electric field (PEF) systems use about 80% less energy than thermal treatments (Kempkes 2010), but are overall more expensive (Montville et al. 2012). PEF technology can be described by high-intensity electric fields which are varied between 20 and 80 kilovolts (kV) per centimeter for a very short time (1–100 ms) (Amiali and Ngadi 2012; Raso et al. 2014). PEFs are able to kill vegetative bacteria through a process known as electropermeabilization (Montville et al. 2012). This process is defined as an electric sock which momentarily opens pores within a bacterial cell's plasma membrane allowing the entry of other macromolecules into the cell, which then leads to cell death. While the mechanism by which PEF kills bacterial spores is not fully elucidated, it is suspected that the mechanism is similar to that of vegetative bacteria (Li and Farid 2016).

In the food industry, a solid food product is passed through PEFs in a continuous system such as on a conveyor belt. The efficacy of the PEF treatment, and its ability to kill microorganisms will depend on the type of food product being tested and depend largely on the type of organism present. It is believed that, in general, Gram-negative bacteria are most susceptible to PEF treatment and bacterial and yeast spores are most resistant (Barbosa-Canovas et al. 1999; Yonemoto et al. 1993). Other factors that affect microbial reductions by PEF include process conditions (field strength, pulse width, pulse frequency, total treatment time, input energy), production conditions (flow rate, holding time, temperature), and food properties (pH, conductivity, particulate) and it is recommended that scale-up and validation studies in a specific PEF system for specific products be performed (Jin et al. 2015). Some argue that high cost is a limitation for commercial use of PEF technology (Kempkes 2010); however, this issue remains controversial.

Initial commercial interests for use of PEF technology mostly surrounded juice and beverage products including tomato juice (Min et al. 2003). In 2016, testing for a pilot-scale batch PEF unit was released for the processing of French fries, potato chips, and other specialty potato products and raw materials including sweet potatoes, cassava, beetroot, and carrots (PotatoPro 2016). With the use of this system, potato cells are electroporated in order to release intracellular compounds, such as reducing sugars involved in the Maillard browning reaction, which reduces the tendency of browning during frying (PotatoPro 2016). This undesirable browning gives the potatoes a burned, or overcooked, appearance. This new innovation could open up the possibilities for the processing of solid plant-based products. Other applications of PEF include mild preservation of beverages and semi-liquid food products, extraction processes such as extraction of antioxidants, extraction of oil and protein from algae, extraction of sugar from sugar beets, and extraction of nutrients or fibers from peels and stems (Pulsemaster 2017). PEF can also be applied for the removal of acrylamide, concentration of protein from potatoes, and enhancement of production processes for cooked ham and dry sausage (Pulsemaster 2017).

5.3.7 Ozone

The antimicrobial effectiveness of ozone has been shown to be much higher than that of chlorine and to affect a broader spectrum of microorganisms than chlorine and other disinfectants (Hirneisen et al. 2010). Bacterial spores have also been shown to be inactivated by ozone (Markland et al. 2013b; Young and Setlow 2004). Studies involving the inactivation of spores by oxidizing agents suggest that inactivation is a result of oxidative damage to the spore's inner membrane (Markland 2011). Ozone can also degrade mycotoxins and pesticides present in foods (Karaca and Velioglu 2007). In addition, there is little concern of residual ozone in treated food products due to the rapid decomposition of ozone into oxygen (Graham 1997) and ozone is currently certified for use on organic foods.

Ozone is applicable in the food industry to treat process water, as a fruit and vegetable wash, in fruit and vegetable storage, and in recycled water (Hirneisen et al. 2010). Aqueous ozone has been used to increase the shelf life of apples, strawberries, and in juices such as apple cider and orange juice (Hirneisen et al. 2010); however, the efficacy of aqueous ozone is largely dependent upon the presence of organic residues, pH, and temperature of the aqueous medium (Hoigné and Bader 1975; Karaca and Velioglu 2007). In general, ozone is more effective at lower temperatures, below pH 5.0, and higher humidity (Karaca and Velioglu 2007).

There are some limitations regarding the use of ozone for food preservation. Ozone is a very reactive molecule that has the ability to inactivate a broad range of microorganisms; however, it also reacts with nearly all organic and inorganic compounds (Karaca and Velioglu 2007). Therefore, the higher the amounts of organic matter present in a food product, the lower the effectiveness of ozone. Ozone may also cause slight deleterious effects on the quality and physiology of food products such as losses in sensory quality including enzymatic browning, antioxidants, vitamins, and minerals (Karaca and Velioglu 2007). Exposure of humans and animals to high levels of ozone can also have detrimental effects on health, which causes concern for workers in processing plants. In the United States, OSHA (Federal Occupational Safety and Health Administration) limits exposure to ozone to a 0.1-ppm threshold for continuous exposure for an 8-h period and 0.3 ppm for a 15-min period (Suslow 2004).

5.4 Conclusions

There are many advantages to the use of advanced processing techniques, yet there are limitations that prevent wider utilization by the food industry. The majority of these technologies are only able to successfully treat food products of a certain moisture content, composition, or surface area. Bacterial sporeformers have also proven to be a challenge for nonthermal technologies as heat is necessary for complete inactivation. More attention should be given to technologies that may prevent the germination of spores. For example, research has shown that when used synergistically, or in various combinations with one another, these technologies as well as the addition of mild heat can effectively inactivate spores. Another limitation regarding these technologies is the lack of consumer acceptance. Consumer education and marketing will be a crucial component as these technologies continue to be developed. Overall, further research should be performed in regard to the advanced processing techniques discussed in this chapter in order to expand their applications. This will allow food companies to be able to continue to provide their consumers with a high variety of products that are safe and shelf-stable.

References

Alvarez, I., B.A. Niemira, X. Fan, and C.H. Sommers. 2006. Inactivation of Salmonella serovars in liquid whole egg by heat following irradiation treatments. *Journal of Food Protection* 69 (9): 2066–2074.

Amiali, M., and M.O. Ngadi. 2012. Microbial decontamination of food by pulsed electric fields. In *Microbial Decontamination in the Food Industry, Novel Methods and Applications*, 407–449. Washington, DC: Woodhead Publishing.

Barbosa-Canovas, G.V., M.M. Gongora-Nieto, U.R. Pothakamury, and B.G. Swanson. 1999. *Preservation of Foods with Pulsed Electric Fields*. San Diego, CA: Academic.

Bialka, Katherine L., Ali Demirci, Paul N. Walker, and Virendra M. Puri. 2008. Pulsed UV-light penetration of characterization and the inactivation of *Escherichia Coli* K12 in solid model systems. *Transactions of the ASABE* 51.

Black, E.P., P. Setlow, A.D. Hocking, C.M. Stewart, A.L. Kelly, and D.G. Hoover. 2007. Response of spores to high-pressure processing. *Comprehensive Reviews in Food Science and Food Safety* 6: 103–119.

Butz, P., and B. Tauscher. 2002. Emerging technologies: Chemical aspects. *Food Research International* 35 (2–3): 279–284.

Chaine, A., C. Levy, B. Lacour, C. Riedel, and F. Carlin. 2012. Decontamination of sugar syrup by pulsed light. *Journal of Food Protection* 75 (5): 913–917.

Chandrapala, Jayani, Christine Oliver, Sandra Kentish, and Ashokkumar Muthupandian. 2012. Ultrasonics in food processing. *Ultrasonics Sonochemistry* 19 (5): 975–983.

Chilton, P., N.S. Isaacs, B. Mackey, and R. Stenning. 1996. The effects of high hydrostatic pressure on Bacteria. In *High Pressure Research in the Biosciences and Biotechnology*, ed. K. Heremans. Belgium: Leuven University Press.

Code of Federal Regulations. 2017. 21 CFR.179: Irradiation in the production, processing, and handling of food. http://www.accessdata.fda.gov/scripts/cdrh/cfdocs/cfcfr/CFRSearch.cfm?fr=179.26. Accessed 30 January 2017.

Delmas, H., and L. Barthe. 2015. Ultrasonic mixing, homogenization, and emulsification in food processing and other applications. In *Power Ultrasonics Applications of High-Intensity Ultrasound*, 757–791. Washington, DC: Woodhead Publishing.

Evelyn, E., and Filipa V. Silva. 2015. Thermosonication versus thermal processing of skim Milk and beef slurry: Modeling the inactivation kinetics of Psychrotrophic Bacillus Cereus spores. *Food Research International* 67: 67–74.

Farkas, J. 2001. In *Food Microbiology: Fundamentals and Frontiers*, ed. M.P. Doyle, L.R. Beuchat, and T.J. Montville, 2nd ed. Washington, DC: Woodhead Publishing.

Gayán, E., I. Álvarez, and S. Condón. 2013. Inactivation for bacterial spores by UV-C light. *Innovative Food Science and Emerging Technologies* 19: 140–145.

Gomez-Lopez, V.M., T. Koutchma, and K. Linden. 2012. Ultraviolet and pulsed light processing of fluid foods. In *Novel Thermal and Non-Thermal Technologies for Fluid Foods*. Cambridge, MA: Academic.

Gould, G.W. 1996. Methods for preservation and extension of shelf life. *International Journal of Food Microbiology* 33 (1): 51–64.

Graham, D.M. 1997. Use of ozone for food processing. *Food Technology* 51 (6): 72–75.

Hijnen, W.A., E.F. Beerendonk, and G.J. Medema. 2006. Inactivation credit of UV radiation for viruses, Bacteria and protozoan OoCysts in water: A review. *Water Research* 40 (1): 3–22.

Hirneisen, K.A., E.P. Black, J.L. Cascarino, V.R. Fino, D.G. Hoover, and K.E. Kniel. 2010. Comprehensive. *Reviews in Food Science and Food Safety* 9 (1): 3–20.

Hoigné, J., and H. Bader. 1975. Ozonation of water: Role of hydroxyl radicals as oxidizing intermediates. *Science* 190 (4216): 782–784.

Iwahashi, H., K. Obuchi, S. Fuji, K. Fujita, and Y. Komatsu. 1996. The reason why Trehalose is more important for Barotolerance than Hasp104 in *Saccharomyces Cerevisiae*. In *High Pressure Research in the Biosciences and Biotechnology*, ed. K. Heremans. Belgium: Leuven University Press.

Jin, T.Z., M. Guo, and H.Q. Zhang. 2015. Upscaling from benchtop processing to industrial scale production: More factors to be considered for pulsed electric field food processing. *Journal of Food Engineering* 146: 72–80.

Karaca, H., and Y. Velioglu. 2007. Ozone applications in fruit and vegetable processing. *Food Reviews International* 23: 91–106.

Keklik, N.M., K. Krishnamurthy, and A. Demirco. 2012. Microbial decontamination of food by ultraviolet (UV) and pulsed UV light. In *Microbial Decontamination in the Food Industry, Novel Methods and Applications*, Woodhead Publishing Series in Food Science, Technology and Nutrition, ed. A. Demirci and M.O. Ngadi, 344–369. Washington, DC: Woodhead Publishing.

Kempkes, M.A. 2010. Pulsed Electric Field (Pef) systems for commercial food and juice processing. In *Case Studies in Novel Food Processing Technologies, Innovations in Processing, Packaging and Predictive Modeling*, Woodhead Publishing Series in Food Science, Technology and Nutrition, ed. C. Doona, K. Kustin, and F.E. Feeherry, 73–102. Washington, DC: Woodhead Publishing.

Knorr, D. 1995. High pressure effects on plant derived foods. In *High Pressure Processing of Foods*, ed. D.A. Ledward, D.E. Johnston, R.G. Earnshaw, and M. Hasting. Nottingham: Nottingham University Press.

Lacombe, A., B.A. Niemira, J.B. Gurtler, X. Fan, J. Sites, G. Boyd, and H. Chen. 2015. Atmospheric cold plasma inactivation of aerobic microorganisms on blueberries and effects on quality attributes. *Food Microbiology* 46 (2015): 479–484.

Li, X., and M. Farid. 2016. A review on recent development in non-conventional food sterilization technologies. *Journal of Food Engineering* 182: 33–45.

Li, H., and M. Gänzle. 2016. Some like it hot: Heat resistance of *Escherichia coli* in food. *Frontiers in Microbiology* 7: 1763.

Lou, F., P. Huang, H. Neetoo, J. Gurtler, B.A. Niemira, H. Chen, X. Jiang, and J. Li. 2012. High pressure inactivation of human norovirus virus-like particles: Evidence that the capsid of human norovirus is highly pressure resistant. *Applied and Environmental Microbiology* 78: 5320–5327.

Markland, S.M. 2011. Characterization of Superdormant Spores of B*acillus Cereus* and B*acillus Weihenstephanensis*. Masters Thesis. Department of Animal and Food Sciences. University of Delaware, Newark DE. http://udspace.udel.edu/handle/19716/11705. Accessed 4 November 2016.

Markland, S.M., D.F. Farkas, K.E. Kniel, and D.G. Hoover. 2013a. Pathogenic Psychrotolerant Sporeformers: An emerging challenge for low-temperature storage of minimally processed foods. *Foodborne Pathogens and Disease* 10 (5): 413–419.

Markland, S.M., K.E. Kniel, P. Setlow, and D.G. Hoover. 2013b. Nonthermal inactivation of heterogeneous and Superdormant spore populations of *Bacillus Cereus* using ozone and high pressure processing. *Innovative Food Science and Emerging Technologies* 19: 44–49.

Mason, T.J., F. Chemat, and M. Ashokkumar. 2015. Power ultrasonics for food processing. In *Power Ultrasonics*, 815–843. Washington, DC: Woodhead Publishing.

Masschelein, Willy J. 2002. In *Ultraviolet Light in Water and Wastewater Sanitation*, ed. Rip G. Rice. Washington, DC: Lewis Publishers.

McDonald, K.F., R.D. Curry, T.E. Clevenger, K. Unklesbay, A. Eisenstrack, J. Golden, and R.D. Morgan. 2000. A comparison of pulsed and continuous ultraviolet light sources for the decontamination of surfaces. *IEEE Transactions on Plasma Science* 28 (5): 1581–1587.

McHugh, T., and B.A. Niemira. 2016. Cold plasma processing of foods. *Food Technology Magazine* 3 (16): 68–72.

Min, S., Z.T. Jin, and Q.H. Zhang. 2003. Commercial scale pulsed electric field processing of tomato juice. *Journal of Agricultural and Food Chemistry* 51 (11): 3338–3344.

Min, S., S.H. Roh, B.A. Niemira, J.E. Sites, G. Boyd, and A. Lacombe. 2016. Dielectric barrier discharge atmospheric cold plasma inhibits *Escherichia coli* O157:H7, *Salmonella*, *Listeria monocytogenes*, and Tulane virus in Romaine lettuce. *International Journal of Food Microbiology* 237 (2016): 114–120.

Montville, Thomas J., Karl R. Matthews, and Kalmia E. Kniel. 2012. *Food Microbiology: An Introduction*. Washington D.C.: ASM Press.

Niemira, B.A. 2014. Irradiation, microwave and alternative energy-based treatments for low water activity foods. In *Microbiological Safety of Low aw Foods and Spices*, ed. M. Doyle, J. Kornacki, and J. Gurtler, 389–401. New York, NY: Springer.

Niemira, B.A., and P. Cooke. 2010. *Escherichia coli* O157:H7 biofilm formation on lettuce and spinach leaf surfaces reduces efficacy of irradiation and sodium hypochlorite washes. *Journal of Food Science* 75 (5): M270–M277.

Niemira, B.A., X. Fan, and C.H. Sommers. 2002. Irradiation temperature influences product quality factors of frozen vegetables and radiation sensitivity of inoculated Listeria monocytogenes. *Journal of Food Protection* 65 (9): 1406–1410.

Niemira, B.A., C.H. Sommers, and G. Boyd. 2003. Effect of freezing, irradiation and frozen storage on survival of Salmonella in concentrated Orange juice. *Journal of Food Protection* 66: 1916–1919.

Niemira, B.A., G. Boyd, and J. Sites. 2014. Cold plasma rapid decontamination of food contact surfaces contaminated with Salmonella biofilms. *Journal of Food Science* 79 (5): M917–M922.

Paidhungat, M., B. Setlow, W.B. Daniels, D. Hoover, E. Papafragkou, and P. Setlow. 2002. Mechanisms of induction of germination of Bacillus subtilis spores by high pressure. *Applied and Environmental Microbiology* 68 (6): 3172–3175.

PotatoPro. 2016. *Pulsed Electric Field for French Fries and Chips: Quantify your benefits with Solidus.* http://www.potatopro.com/news/2016/pulsed-electric-field-french-fries-and-chips-quantify-your-benefits-solidus. Accessed 7 November 2016.

Pulsemaster. 2017. FAQ about pulsed electric field processing: What are typical applications of PEF processing? https://www.pulsemaster.us/pef-pulsemaster/faq. Accessed 13 February 2017.

Raso, J., M.M. Gongora-Neito, G.V. Barbosa-Canovas, and B.G. Swanson. 1998. Influence of several environmental factors on the initiation of germination and inactivation of *Bacillus Cereus* by high hydrostatic pressure. *International Journal of Food Microbiology* 44 (1–2): 125–132.

Raso, J., S. Condon, and I. Alvarez. 2014. Non-thermal processing: Pulsed electric field. In *Encyclopedia of Food Microbiology*, ed. C.A. Bratt, 966–973. New York: Academic.

San Martin, M.F., G.V. Barbosa-Canovas, and B.G. Swanson. 2002. Food processing by high hydrostatic pressure. *Critical Reviews in Food Science and Nutrition* 46 (6): 627–645.

Setlow, P. 2003. Spore germination. *Current Opinion in Microbiology* 6 (6): 550–556.

Suslow, Trevor V. 2004. Ozone applications for postharvest disinfection of edible horticultural crops. *University of California ANR*, 1–8.

Szent-Gyorgyi, A. 1933. Chemical and biological effects of ultra-sonic radiation. *Nature* 131 (3304): 278.

United States Department of Agriculture. 2013. "Danger Zone". http://www.fsis.usda.gov/wps/portal/fsis/topics/food-safety-education/get-answers/food-safety-fact-sheets/safe-food-handling/danger-zone-40-f-140-f/CT_Index. Accessed 4 November 2016.

Wuytack, E.Y., J. Soons, F. Poschet, and C.W. Michiels. 2000. Comparative study of pressure- and nutrient-induced germination of Bacillus subtilis spores. *Applied and Environmental Microbiology* 66 (1): 257–261.

Yonemoto, Yoshimasa, Tetsuo Yamashita, Masafumi Muraji, Wataru Tatebe, Hiroshi Ooshima, Jyoji Kato, and Akira Kimura. 1993. Resistant of yeast and bacterial spores to high voltage electric pulses. *Journal of Fermentation and Bioengineering* 75 (1): 99–102.

Young, S.B., and P. Setlow. 2004. Mechanisms of *Bacillus Subtilis* spore resistance to and killing by aqueous ozone. *Journal of Applied Microbiology* 96 (5): 1133–1142.

Chapter 6
Packaging of Perishable Food Products

Cynthia Ebner, Angela Morgan, and Clyde Manuel

6.1 Introduction

While often overlooked or simply disregarded as waste, packaging is an integral component of the modern-day food production system, allowing for the benefits of food processing to be maintained long after production. Without food packaging, modern food products would have limited shelf life, be susceptible to potentially hazardous and deleterious contamination, and would lack numerous features designed for convenience that many consumers have grown accustomed to. With these considerations in mind, it is easy to appreciate the important role of packaging in delivering food of acceptable quality to the consumer.

In this chapter, the reader is presented with a broad overview of the world of food packaging, with a specific emphasis on content relevant to the packaging of perishable food products, such as meats, cheeses, and fresh produce. This chapter is not intended to be a comprehensive study of all aspects of food packaging, as there are numerous excellent reviews that fit this role (Del Nobile and Conte 2013; Robertson 2012).

C. Ebner
Sealed Air Corporation, Charlotte, NC, USA

A. Morgan
Aptar Group, Crystal Lake, IL, USA

C. Manuel (✉)
GOJO Industries, Akron, OH, USA
e-mail: manuelc@GOJO.com

© Springer Nature Switzerland AG 2021
P. J. Taormina, M. D. Hardin (eds.), *Food Safety and Quality-Based Shelf Life of Perishable Foods*, Food Microbiology and Food Safety,
https://doi.org/10.1007/978-3-030-54375-4_6

6.2 Packaging Functions

6.2.1 Containment

Quite possibly, the most basic function of a package, containment, allows for the transport and storage of food and other goods. At first glance, it may seem easy to dismiss this function of a package as blatantly obvious, but its importance is nevertheless acknowledged; simply put, a package will not function properly if it does not contain its contents. Some of the earliest forms of packaging, such as pottery and bags, were originally intended for the containment of food. In one instance, chemical analysis of the lipid content on ancient pottery from Japan suggests that ceramic pots were used to store, transport, and process fish and seafood as far back as 15,000 years ago, many years before the advent of farming practices (Lucquin et al. 2016). The time frame of this discovery suggests that the widespread adoption of pottery may not have coincided with the settling of mankind due to farming, but in fact may have arisen much earlier during the age of hunter gatherers.

6.2.2 Protection

Perhaps the most important function of a food package is to protect its contents from becoming unfit for consumption. A product may become unfit for consumption if its perceived quality falls below a level of consumer acceptance, or if it is no longer safe to consume. In general, factors that contribute to the loss of product acceptability fall into one of three categories: physical, chemical, and biological. Packaging can play a major role in preventing product deterioration from each of these three factors.

Food products are subjected to physical abuse throughout their entire life cycle, including production, transit, and even during storage and display on store shelves. A damaged package will lead to loss of quality and reduced shelf life of a product, as the damaged package is no longer able to protect the food product from the elements. Because of this, great care is often taken to design a packaging system that minimizes the effects of physical abuse. On the production floor, a packaged product will have to withstand the abuse of both manual and automatic handling. For example, it is not uncommon for vacuum packaged meat products to puncture while being transported on conveyor systems, especially if the system is in a state of disrepair or if processing and packaging line speeds are exceedingly fast. To combat this, packaging materials used for vacuum packaged meat products are sometimes designed with heavier gauge materials that are more resistant to abuse due to their increased thickness. During transportation, products typically experience mechanical shock in the forms of drops and vibrations. Materials such as corrugated paperboard are typically used for building pallets for shipment purposes, as it offers exceptional crush protection.

Many food products can become unacceptable as the result of degradative chemical processes. Oxygen, moisture, and light are the primary extrinsic factors that can initiate and accelerate these processes, and the material and type of food packaging can be tailored to minimize these chemical processes. Meat is often vacuum packaged in materials with oxygen barrier properties in order to slow the progression of rancidification, which occurs when oxygen accelerates the breakdown of unsaturated fatty acids into volatile aldehydes and ketones, which emit an unpleasant and rancid odor. Moisture gain or loss is another factor that contributes to the loss of quality of food products. Products with extremely low water moisture contents, such as crackers or potato chips, are often packaged in materials that have moisture barrier properties in order to prevent staling caused by an influx of moisture into the package. Finally, excessive light exposure can reduce the shelf life of many foods, especially those with a high fatty acid content. Beer is a product familiar to consumers that is particularly susceptible to the deleterious effects of light. Packaging of beer in metal kegs, cans, or brown bottles, has been shown to significantly increase the shelf life of beer beyond that of clear packages, since the rate of photolysis of hop alpha acids is reduced in these packages (Heyerick et al. 2003). Beer that has undergone photolysis of hop alpha acids is easily recognizable by the consumer as having a cardboard or skunky taste.

Packaging can also protect food from biological spoilage. Pests such as insects and rodents can eat through packaging materials and compromise the contents of the package. This can be prevented by packaging food in materials that are pest-resistant (often by increasing material thickness), storing palletized products properly in warehouses, and utilizing pest control solutions within the storage environment. Spoilage and pathogenic microorganisms can render a food inedible and even unsafe if allowed to grow to high levels, and packaging can serve as a barrier to entry for these microorganisms. The foodborne pathogen, *Listeria monocytogenes,* for example, is a contaminant in the environment of some food processing plants, especially those that produce ready-to-eat (RTE) foods including meat products. As an additional *L. monocytogenes* control measure, many RTE meat processors apply a post-packaging lethality step on the finished product by hot water immersion. This involves the use of a higher heat-resistant type of packaging film and results in both a reduction in pathogen numbers and extended shelf life for the product.

6.2.3 Convenience

Many design aspects of food packages have arisen from the need for added convenience throughout the life cycle of a product. For example, a product's distribution chain can be optimized and streamlined when the product is distributed by organizing finished product into units at varying steps. At the primary level, products are individually packaged for display purposes and for consumer purchase. At the secondary level, multiple units destined for product display are organized into a large package (typically made of corrugated paperboard case) that adds convenience when transporting products at the store level. Finally, at the tertiary level, secondary units of products are organized together for the purposes of shipment, usually on a stretch-wrapped pallet. Organizing products into these units simplifies the logistics involved in transporting many products over long distances.

In recent years, numerous commercial examples of innovative convenience features have emerged with direct ties to the shelf life of a food product. In many cases, these features fall under the category of intelligent packaging, which is defined as packaging that can track, sense, and/or measure some aspect of the contained product and then communicate this information to the consumer or user. While still in their commercial infancy, these intelligent packages can communicate the shelf life status to the consumer. Intelligent packaging is discussed in more detail later in this chapter.

6.2.4 Communication

The package that encloses food also provides a platform for communicating important information to the consumer. Food companies can differentiate themselves from their competitors by incorporating marketing or branding materials into their package designs. In the United States, the Food and Drug Administration (FDA) regulates much of the information that appears on the outside of a primary food package intended for consumer purchase. Information such as statements of identity, nutritional facts, ingredient declarations, package contents, net weight, and serving sizes are examples of information required by law. These regulations are designed to assist consumers in making informed decisions on their food purchases. It should be noted that messages tied to product shelf life, such as "use by," "best before," "sell by," or "expired by" dates, are not regulated and thus not required by law (the sole exception to this being infant formula).

6.3 Materials Used in Food Packaging

Food can be packaged in a wide variety of materials. The ideal choice of material for a particular food product is influenced by a variety of factors, including cost, appearance, flexibility, durability, ease of implementation with production process, and compatibility with food. For perishable products such as packaged meat, seafood, and fruits and vegetables, one of the most important variables to consider is permeability of the material to moisture, oxygen, carbon dioxide, and also light, as these factors are highly influential on various processes that dictate the shelf life of a food product. These factors are not discussed at length in this chapter but are discussed in general terms.

6.3.1   Metal

Metals have a variety of advantages as a packaging material. They offer superior physical durability, barrier protection from gas transmission and light penetration, and are highly recyclable. Steel and aluminum are the two most frequently used metals for food packaging. Most cans produced in the United States each year are made of steel, which is typically coated with a thin layer of electrically deposited tin to enhance corrosion resistance (hence the name "tin can"). Aluminum, while more expensive than steel, offers advantages in the form of lighter weight and enhanced corrosion resistance. Most metal containers used for food products are lined with an inert protective enamel coating that prevents contact between the product and the metal itself. This is especially important for high acid foods such as soda, juices, and tomato-based products, as contact with the metal itself will rapidly leach metallic ions into the food itself. The leaching of metallic ions into the food leads to a reduction in product quality due to both flavor loss and nutrient loss. A well-known example of this is the rapid loss of ascorbic acid (vitamin C) in the presence of metallic ions. In addition to rigid containers such as cans and trays, metallic layers can also be incorporated into flexible packages by lamination or by metalizing plastic film in a vacuum chamber, thus incorporating a high-quality barrier into a flexible package. Metal packaging is most often used for shelf stable products, as its properties lend itself well to the harsh processing conditions these products undergo.

6.3.2   Glass

Glass is a non-crystalline amorphous solid derived from the heating of silica oxides in the presence of various additives. A number of additives in the form of carbonates are typically used in the glass-making process depending on the desired effect on the finished product. For example, sodium and potassium carbonate are often added to the glass formula in order to lower the melt temperature, making the glass easier to work with during the manufacturing process. During the manufacturing process, furnaces heat the raw material mixture to approximately 1500 °C, a temperature at which the raw materials melt into a viscous liquid and can then be molded into shapes. Upon cooling, the molds harden into a non-crystalline amorphous solid. Glass is a highly recyclable material, with recycled broken glass (cullet) constituting a large proportion (15–50%) of the ingredients used in a formulation for new glassware. The advantages of glass include its ability to be formed into a variety of shapes, its impermeability to moisture and gases, and its non-reactivity with food. These factors make glass an ideal packaging material for long-term storage of products that might be susceptible to flavor loss. Some disadvantages of glass include heavy weight (which increases shipping costs) and breakability from physical damage or rapid temperature fluctuations. Consumers often interpret glass packaging as an indicator of higher product quality, especially for dressings, sodas, and juices (Risvik 2001).

6.3.3 Plastics

The word *plastic* describes any material that can be molded into various shapes and forms while soft and then set into a rigid or flexible form. Plastics can be categorized as thermoset or thermoplastic materials. Thermoset materials solidify into an irreversible rigid state, usually offering exceptional durability at the expense of reduced recyclability. On the other hand, thermoplastic materials soften when heat is applied. Nearly all plastics used in food packaging are considered thermoplastic. Within the context of food packaging, the term *plastic* typically describes synthetic materials derived from petroleum by-products such as ethylene and methane, although natural materials such as cellulose and lactic acid are sometimes used. Plastic materials are created through a process called polymerization, which links together individual monomeric units together to create higher molecular weight polymeric materials. This is the reason the term *plastic* and *polymer* are sometimes used interchangeably, although this can be a source of confusion.

Plastics have numerous advantages as a material for food packaging. They are relatively inexpensive, lightweight, and can be molded into a variety of shapes, which can help to reduce costs associated with transportation of the material. Owing to its flexibility, plastics are also highly resistant to denting and shattering (though they are highly susceptible to puncturing). Plastics are also extremely versatile materials, and the polymerization process allows for the creation of a wide variety of materials with various functions and properties depending on the end goal. The major disadvantage of plastics is their relatively poor recyclability when compared to other materials, such as glass or metals. This is especially true for more complicated plastic food packaging materials, such as multilayered laminates or co-extruded materials.

6.4   Specific Plastic Polymers Used in Food Packaging

6.4.1   Polyethylene

Polyethylene is created from the polymerization of ethylene, a gaseous by-product from the petroleum industry. It is the most abundant plastic material produced, with an average annual global production of ethylene resin around 80–90 million metric tons per year (Strom and Rasmussen 2011). It has a variety of advantages that make it useful for food packaging, including ease of production, low cost, excellent formability, strength, good resistance to moisture and chemicals, and highly recyclable.

Polyethylene materials can be classified into functional groups based on their polymer chain branching and density. Low-density polyethylene (LDPE) is a highly branched polymer, which interferes with the ability of the polymer to stack tightly into itself, thus decreasing its density (Fig. 6.1). The low density of LDPE results in the material having a low melting temperature, which makes it a good heat sealable material. LDPE is also relatively transparent with good optical properties and good moisture resistance. These favorable properties make LDPE frequently used in packaging applications that require a film structure.

Linear low-density polyethylene (LLDPE) has less branching than LDPE, resulting in a relatively linear polymer (Fig. 6.1). This linearity results from its unique manufacturing process, which uses butene, hexene, and octene in the copolymerization process. The resultant polymer has a much narrower melt temperature range than typical LDPE, and also higher tensile strength and impact resistance. This allows for a thinner gauge film to be used in many applications. Disadvantages of LLDPE include poorer optical properties than LDPE and difficulty of manufacture and processing.

High-density polyethylene (HDPE) is a linear polymer with a higher density than LDPE or LLDPE. The primary advantage of HDPE over LDPE is enhanced durability due to its high tensile strength. It also has a higher melting temperature and increased resistance to cracking at lower temperatures. HDPE has poor optical properties, and so is rarely used in applications requiring package clarity. Plastic milk containers are one of the best-known examples of a food package based on HDPE. Additional examples of objects comprised of HDPE include bottle caps, industrial piping, and in some instances, even fuel cells for automobiles.

Fig. 6.1 Depiction of the general linearity properties of HDPE, LLDPE, and LDPE

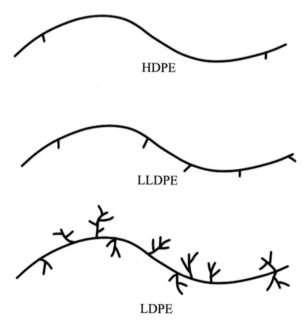

6.4.2 *Polypropylene*

Polypropylene (PP) is a plastic material made from the catalytic addition of propylene monomers into a polymer (Fig. 6.2). Depending on the catalyst conditions, the resultant polypropylene polymer can take one of three configurations (Fig. 6.3): isotactic (methyl groups are on one side of the carbon chain), syndiotactic (methyl groups are evenly dispersed on both sides of the carbon chain), and atactic (methyl groups are randomly dispersed on both sides of the carbon chain). Atactic PP is a low-quality and low-value by-product of the PP polymerization process, as it is a soft, rubbery material with a lower melting temperature than that of isotactic or syndiotactic PP. Commercial PP is primarily in the isotactic and syndiotactic forms, as these are more crystalline and predictable than atactic. PP is an extremely tough and dense polymer, with enhanced resistance to chemicals and heat. Its high melting temperature makes it a good polymer for use in such applications as hot filled products, retort pouches, and microwavable containers. Unlike the polyethylenes, PP has poor transparency, and appears as a hazy material when compared to other polymers. Common examples of PP in food packaging include rigid trays and food storage containers.

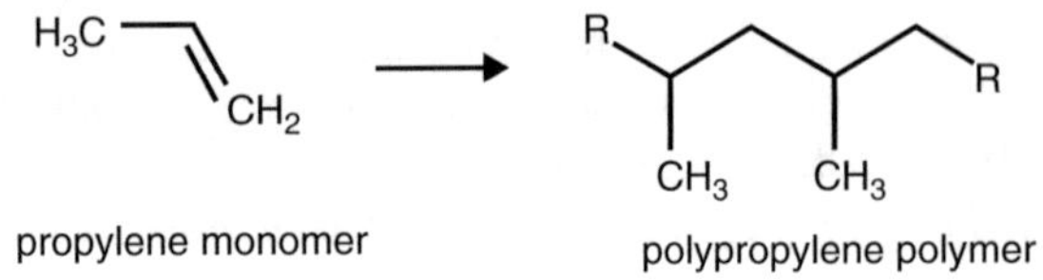

Fig. 6.2 Polymerization of propylene into polypropylene

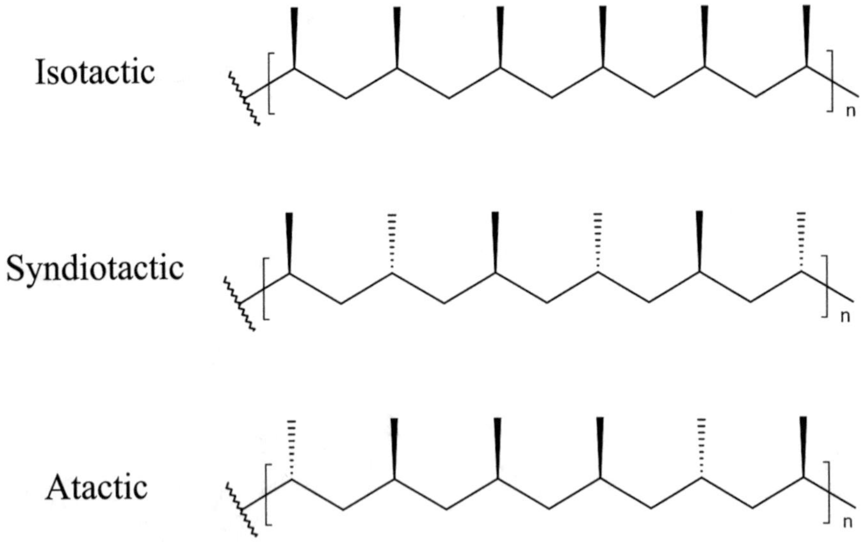

propylene monomer

polypropylene polymer

Isotactic

Syndiotactic

Atactic

Fig. 6.3 Various configurations of polypropylene and their structures

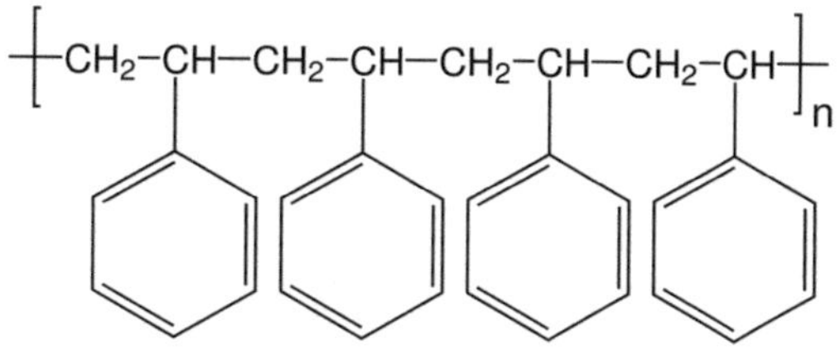

Fig. 6.4 Chemical
structure of polystyrene

$$\left[CH_2-\underset{\displaystyle \bigcirc}{CH}-CH_2-\underset{\displaystyle \bigcirc}{CH}-CH_2-\underset{\displaystyle \bigcirc}{CH}-CH_2-\underset{\displaystyle \bigcirc}{CH} \right]_n$$

6.4.3 Polystyrene

Polystyrene (PS) is a material made from the addition of styrene monomers. The resulting polymeric chain has a benzene ring attached at every other carbon (Fig. 6.4). PS is rather inexpensive but suffers from several properties that make it impractical for use in the packaging of perishable foods. PS is a very brittle material, making it a poor choice when mechanical stability is required. PS is also a poor barrier to moisture and oxygen, meaning it is not practical for use with technologies such as modified atmosphere packaging. PS can be stretched during the extrusion process, resulting in a product called oriented PS film. Oriented PS film has improved optical properties but remains extremely brittle. This form of PS is sometimes used as an inexpensive alternative to PP trays, especially for packaging of meats. Expanded PS (EPS) is an extremely common material with insulative properties. EPS is produced by the addition of a blowing agent, usually pentane, into EPS resin. When heat is applied to the EPS resin beads, the beads expand to several times their original size. The beads are then molded together into a closed-cell foam with excellent insulative properties. Typical food applications for EPS are egg trays, meat trays, and coffee cups. In North America, consumers generally refer to EPS as "Styrofoam." This is technically incorrect, as the name "Styrofoam" is owned and trademarked by the Dow Chemical Company and is used as a building insulation material.

6.4.4 Polyvinyl Chloride

Polyvinyl chloride (PVC) is a polymer produced from the addition of vinyl chloride monomers (Fig. 6.5). This plastic material offers good optical and strength qualities while being resistant to oils and other hydrophobic compounds. PVC is prone to degradation at high temperatures, and therefore most PVC is produced using plasticizers in the production process in order to lower the melting temperature of the product. PVC films produced using these plasticizers often have very good stretch properties, making them excellent materials for overwrapping some food products, such as meat.

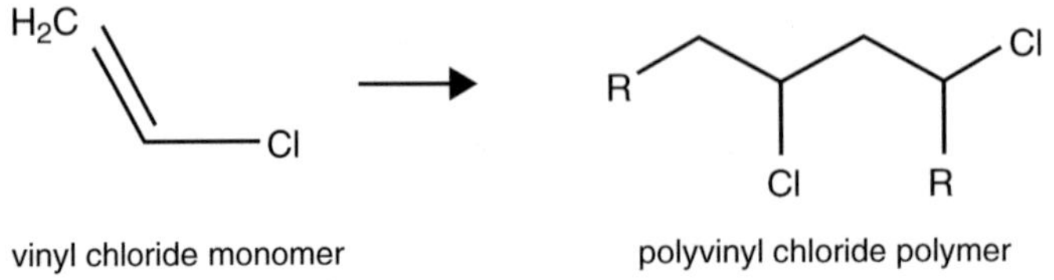

vinyl chloride monomer   polyvinyl chloride polymer

Fig. 6.5 Polymerization of vinyl chloride into polyvinyl chloride

The use of plasticizers in the production of PVC (and other plastic materials) has been highly controversial. Recent studies have demonstrated detectable levels of bisphenol-A (BPA) in the urine of a reference population of adults in the United States (Calafat et al. 2005). This is alarming as BPA is a known carcinogen with endocrine-disrupting properties and has been associated with an increased risk of developing breast and prostate cancers in mammals (Seachrist et al. 2016). While there has yet to be a scientific consensus on the risks of BPA exposure in food packaging materials, many states in the United States have already banned the use of BPA in products intended for use in infants and children. At the federal level, The National Institute of Environmental Health Sciences (NIEHS) and National Toxicology Program (NTP) have recently funded over $30 million in grants for research to further investigate the risks of BPA exposure (Schug et al. 2013).

6.4.5 Polyvinylidene Chloride

Polyvinylidene chloride (PVDC) is similar in structure to PVC, except that it has two chlorine atoms per monomeric unit. Commercial polymerizations of PVDC typically include a co-monomer to lower the melt temperature of the material. The major advantage over PVDC over PVC is its excellent moisture and gas barrier properties. Due to high costs, PVDC is often used as an individual component in more complicated multilayered packaging materials, for example, in some barrier shrink bags for vacuum packaging beef. This material is frequently sold to households as a film wrap for food. Saran® Wrap is an example of a popular brand of PVDC film wrap used for food. Originally created by the Dow Chemical Company, this PVDC-based film wrap was first marketed to households in 1953. The SC Johnson Company purchased the rights to Saran® Wrap in 1998, and shortly thereafter reformulated the product to be based on polyethylene.

6.4.6 Ethylene Vinyl Alcohol and Polyvinyl Alcohol

Ethylene vinyl alcohol (EVOH) and Polyvinyl alcohol (PVOH) are two polymers frequently used in multilayered packaging applications for their excellent gas barrier properties. This barrier function is due to the OH group contained in their monomeric units. Two main disadvantages of these materials are their high cost and high solubility in water. For this reason, these two polymers are usually used in combination with other polymers in a multilayered package, in order to protect the EVOH or PVOH from hydrolysis. EVOH is the most frequently used polymer for oxygen barrier properties in multilayered packaging systems.

6.4.7 Polyethylene Terephthalate

Polyethylene terephthalate (PET) is one of the most widely used polymers for food packaging materials. This material is commonly produced as a product of a trans-esterification reaction with ethylene glycol and dimethyl terephthalate, or by an esterification reaction with ethylene glycol and terephthalic acid. PET is a member of the "polyester" family of materials, which means that it includes an ester functional group in its polymer chain. The term polyester is commonly used to describe PET materials. The material has relatively good optical and barrier properties, is stable over a wide temperature range, and has excellent mechanical and chemical resistance. This makes PET an excellent material for a variety of food packaging applications, especially those that are exposed to extreme temperature ranges.

6.4.8 Polyamides

Polyamides are a wide group of polymers that incorporate an amide group into its polymeric backbone. In the United States, these materials are commonly referred to as nylon materials, although this term is technically a former DuPont trademark. Polyamide materials are resistant to high temperatures and mechanical stress. They are relatively good gas and odor barriers, but often have poor water barrier properties. An advantage of polyamide as a barrier material over EVOH is that its barrier performance is not typically impacted by moisture content. Polyamide materials are extensively used in the packaging of cheese and dairy products and cured meat products.

6.4.9 Polycarbonates

Polycarbonates (PC) are polymers containing carbonate groups in its backbone. They are formed by the polymerization of Bisphenol-A (BPA) with phosgene. The resultant material is clear, extremely durable, and very resistant to heat. This material is commonly used in the manufacture of refillable water containers, frozen food trays designed for oven reheating, and some carbonated beverages. Despite the advantages of polycarbonates as a packaging material, the use of BPA in the polymerization process is highly controversial (see section above on polyvinyl chloride).

6.4.10   Ionomers

Ionomers are copolymers of ethylene and methacrylic or acrylic acid that is partially neutralized by metal cations, typically zinc or sodium. Ionomers often have a highly desirable balance of properties, including excellent optical clarity, high puncture resistance, enhanced durability and toughness (relatively to PE films), low heat seal initiation temperatures, broad seal strength over a wide variety of temperatures, and excellent resistance to oils and lipids. Ionomers are often used in multilayered coextruded or laminated packaging materials as tie layers, facilitating adhesion of two relatively incompatible materials.

6.5   Packaging of Specific Perishable Food Products

6.5.1   Overview

In the following section, specific examples of packaging technologies and their application to extending the shelf life of perishable food products are considered. Prior to their discussion, it is first important to understand the main packaging technologies used to extend the shelf life of these products: vacuum packaging and modified atmosphere packaging (MAP).

6.5.2   Vacuum Packaging

Initially developed by Henri de Poix of the Dewey Almy Chemical Company prior to World War II, the process was finally patented in 1945, and involved storing frozen meat quarters in a latex rubber bag that was subjected to a vacuum (De Poix 1945). While today's version of vacuum packaging relies on plastic polymer materials instead of latex rubber, the fundamental process remains the same: (a) the food product is first placed into a flexible and impermeable, plastic bag or container; (b) the atmosphere in the package is removed by evacuation; and (c) the plastic bag or container is hermetically sealed to prevent transmission of moisture or gases into the sealed package. In vacuum packaging, the objective is to remove as much oxygen as possible within the package. In general, modern vacuum packaging technologies are able to achieve oxygen levels of around 0.5–1.5% at the point of packaging (Kelly et al. 2018).

6.5.3   Modified Atmosphere Packaging

Modified atmosphere packaging (MAP) is a process where a mixture of gases, typically purified nitrogen, oxygen, or carbon dioxide, is flushed into a package just prior to the sealing step. Packages used in a MAP process typically consist of a rigid tray with an impermeable or semi-permeable multi-layered barrier film on top. On packages with an impermeable barrier, the gas mixture chosen remains constant throughout the life of the product. Semi-permeable layers allow for gas exchange between the package and the environment, which is critical for fruits and vegetables that respire long after harvest.

The mixture of gases chosen for MAP depends upon the desired effect. Oxygen is often added to prevent creating a strict anaerobic environment, which may favor the proliferation of spoilage and/or pathogenic bacteria in some foodstuffs (Farber 2016). Carbon dioxide is utilized for its antimicrobial effect, as it is able to easily penetrate into bacterial cells, causing cell death via cell wall collapse and subsequent leaking of cellular contents (Oulé et al. 2006). Carbon dioxide is also is more soluble than oxygen in water, fats, and oils, and thus has a limited ability to penetrate into the surface of foods. Nitrogen is solely utilized as inert filler, as it displaces other reactive gases. Less common gases utilized in MAP technology include carbon monoxide and sulfur dioxide. In meat products, carbon monoxide can be used to permanently "fix" myoglobin into the carboxymyoglobin state, which then appears as a bright red pigment (Djenane and Roncalés 2018). However, this practice is highly controversial due to issues with consumer acceptance. Carbon monoxide can also be used to prevent the browning of fruits and vegetables. Sulfur dioxide is sometimes used as an antimicrobial, especially in preventing mold and bacteria from growing on fruits; however, it can cause allergy-like reactions in some sensitive consumers and therefore is rarely used (Thompson 2010).

6.5.4 Packaging of Fresh Red Meat

Research on consumer preference of fresh red meat has demonstrated color to be the primary factor influencing consumer purchasing decisions, with consumers having a preference for red meat with a bright cherry red color (Font-i-Furnols and Guerrero 2014). Myoglobin, a meat pigment found in the muscle tissue of animals, is responsible for the familiar red color of beef. Myoglobin is a protein that stores oxygen in the muscle tissue of animals. This function is attributed to a heme group within myoglobin, which has a strong affinity for iron and oxygen. In meat, the color is largely determined by the oxidation state of the myoglobin molecule, which can be myoglobin (purple), oxymyoglobin (bright red), or metmyoglobin (brown). In the absence of oxygen, the myoglobin (purple) state predominates with the heme molecule containing an iron atom in the ferrous (+2) oxidation state bound to a water molecule (H_2O). When exposed to oxygen for a short period of time, the myoglobin state changes to oxymyoglobin (bright red) where the heme molecule contains an iron atom in the ferrous (+2) oxidation state bound to an oxygen molecule, resulting in the familiar "bloom" to consumers. When fresh meat has been exposed to oxygen for a long period of time, or when it has been cooked, it will convert to the metmyoglobin state (brown). This is when the iron atom in the heme molecule is in the ferric (+3) state, as it has lost an electron. Meat color can also be impacted by age of the animal at slaughter, animal diet, and level of exercise. Since the metmyoglobin (brown) state of meat is seen as a sign of poor quality (even though it is not associated with reduced quality from a microbial standpoint), most efforts to extend the shelf life of red meat products seek to delay or prevent the formation of metmyoglobin.

Microbial spoilage is also a major factor in influencing the shelf life of fresh red meat. Given that essentially all packaged meat products are produced, shipped, and stored under refrigeration conditions, the predominant spoilage microflora are psy-chrotrophs, most notably *Pseudomonas* species in aerobic packaged meat, and *Lactobacillus* and *Brochothrix* species in meat packaged under reduced oxygen conditions. When allowed to flourish on meat, *Pseudomonas* species produce pro-teolytic enzymes that break down amino acids into foul smelling volatile com-pounds, appearing as highly offensive off odors to the consumer. *Pseudomonas* species can also cause slime to develop on meat as a result of tissue proteolysis. This slime often appears on spoiled meat as a green discoloration.

Given the severity of proteolytic spoilage and reduced shelf life associated with more aerobic type packaging systems, the vast majority of packaging solutions that seek to extend the shelf life of fresh red meat are focused on reducing the oxygen content of the package itself. Vacuum packaging is by far the most common method of choice for shipment and delivery of fresh red meat primals and subprimals, and can achieve a shelf life of approximately 7–12 weeks (Voges et al. 2007). This shelf life can vary depending upon temperature conditions, beef source, and initial microbial load. At the store, the vacuum packages are then opened and broken down into indi-vidual steaks by the butcher. The individual cuts are placed onto a foam tray with a moisture absorbent pad wrapped with a clear oxygen and moisture permeable plastic film. The high permeability of the plastic overwrap material allows for oxygen to come into contact with the meat, producing the bright cherry red "bloom" that consumers prefer. While this approach for retail packaging of beef is very inexpensive, it also results in a short shelf life of approximately 4–7 days, depending on storage conditions and initial beef quality. Vacuum packaging of individual steaks and cuts is possible and is sometimes seen; however, consumers tend to find the purple color of the native myoglobin found in vacuum packaged beef unacceptable (Voges et al. 2007).

Innovations in packaging technology have resulted in fresh red meat products being packaged under MAP conditions. These packages typically consist of meat placed inside a semi-rigid barrier tray and sealed with a barrier top web. An advan-tage of this approach is that the product is not physically compressed, as can happen at the corners of vacuum bags. This compression can result in exudation of cellular components from the muscle tissue itself, also known as "purge," which many con-sumers find undesirable. Gas mixtures for red meat packaged using MAP technol-ogy usually contain high amounts of oxygen (50–80%), with carbon dioxide as the remainder. The high oxygen content helps to ensure that the muscle tissue remains a bright red color, while the carbon dioxide serves to help control spoilage microor-ganisms. An unintended consequence of the high oxygen content in these packages is the tendency for the meat to spoil due to lipid oxidation, usually after a period of around 2 weeks. Low-oxygen MAP packaging generally uses gas mixtures of around 30% carbon dioxide and 70% nitrogen. The shelf life of refrigerated red meat packaged under these conditions can range anywhere from 3 to 5 weeks. A disadvantage is that, depending on the gaseous mixture of the MAP system, these conditions favor the purple color of the native myoglobin form in the muscle tissue, which some consumers may not prefer. Another MAP system commonly used for case-ready fresh meats is the master pack system that consists of several air permeable overwrapped packaged placed in a large pouch commonly referred to as a master pack or mother bag. The master pack is impermeable to oxygen and moisture. The retailer then removes the oxygen permeable case-ready packages from the high barrier master package and the product blooms within several minutes after exposure to air.

6.5.5 Packaging of Poultry Products

Unlike beef, the shelf life of poultry products is primarily tied to the microbial production of off flavors and odors, and not to color. The microorganisms primarily responsible for these off flavors and odors are *Pseudomonas, Achromobacter, Enterobacter*, and *Shewanella* species (Russell et al. 1995). In a sensory panel on expired poultry products, consumers described off flavors from these microorganisms as "sulfur," "dishrag," "ammonia," "wet dog," "skunk," "dirty socks," and "rancid fish" (Russell et al. 1995). Poultry products are typically packaged under MAP conditions more frequently than vacuum conditions. Gas mixtures for poultry products packaged under MAP conditions can vary greatly, though in general, researchers have identified better shelf life performance when packaged at carbon dioxide levels of 20% or higher in order to prevent proliferation of aerobic spoilage bacteria (Rossaint et al. 2014). Under these conditions, a shelf life of 23 weeks is typically achievable. The majority of poultry retail packages consist of a rigid or foam tray sealed by a barrier overwrap film. This allows for the package to be gas flushed to extend shelf life, all while maintaining the traditional look of a retail poultry overwrap package.

A major safety consideration of MAP packaged poultry concerns the growth of *Campylobacter jejuni*, a mesophilic pathogenic organism that thrives in low oxygen environments. Compounding the issue is the widespread distribution of *Campylobacter jejuni* in retail poultry products within the United States. In one survey, 70.7% of chicken samples tested at retail locations tested positive for *Campylobacter* (Zhao et al. 2001). Thus, temperature control is critical for poultry products packaged using MAP gas mixtures that may favor growth of this pathogenic organism.

6.5.6 Packaging of Seafood Products

Seafood products such as freshwater fish, saltwater fish, and shellfish are extremely perishable products with spoilage resulting from both autolytic changes involving proteolytic enzymes as well as from microbial degradation. Immediately following the death of a fish, endogenous enzymes digest muscle tissue into compounds with off-odor properties (Ahmed et al. 2013). These endogenous enzymes are able to hydrolyze protein tissue even at refrigeration temperatures. After the autolysis process has continued for 3–5 days, these compounds begin to serve as nutrients for a variety of spoilage bacteria, which then proliferate and produce their own undesirable odors and off flavors. Typical spoilage organisms for seafood include *Pseudomonas* spp., *Shewanella* spp., *Photobacterium* spp., *Brochothrix* spp., and various lactic acid bacteria (Boziaris and Parlapani 2017). Most seafood is also highly susceptible to chemical spoilage, primarily in the form of lipid oxidation. This problem is compounded by the fact that these oxidation reactions can occur at freezing temperatures. Also, most seafood products are comprised of a high percent of unsaturated fatty acids, which are highly susceptible to oxidation.

Vacuum and MAP packaging can extend the shelf life of seafood products by shifting the microflora towards non-spoilage organisms. Early research has shown that high carbon dioxide environments approaching 100% could vastly improve on the shelf life of seafood products, albeit at the expense of package collapse, a phenomenon that results from the dissolution of atmospheric carbon dioxide into the muscle tissue of the seafood (DeWitt and Oliveira 2016). The gas compositions used in MAP packaging of fish and seafood generally fall into one of two categories: mixtures for fatty fish (a mixture of carbon dioxide and nitrogen) and mixtures for non-fatty fish (an even mixture of oxygen, carbon dioxide, and nitrogen). The shelf life of packaged fish products is generally between 10 and 20 days, depending on the storage temperature and incoming quality of the initial product.

A major risk associated with vacuum and MAP packaged fish is the potential growth of *Clostridium botulinum* type E, an anaerobic organism that produces the potentially deadly botulinum toxin. This risk stems from observations that many fish may become naturally contaminated with type E strains of *Clostridium botulinum* that reside in seabeds (Gram 2001). Originally, oxygen was added to fresh fish packages in order to prevent the growth of *Clostridium botulinum*, but it has now been shown that this organism can grow and produce toxin in packages with relatively high levels of oxygen added. Exacerbating this issue is the observation that several strains of type E *Clostridium botulinum* can produce botulinum toxin at refrigeration temperatures that normally suppress spoilage organisms (Gram 2001). Thus, a packaged fish product that undergoes temperature abuse could readily develop toxic levels of botulinum toxin without any glaring effects of spoilage organisms. The only effective way to prevent growth of *Clostridium botulinum* in vacuum or MAP packaged fish is to keep the product at or below 3 °C at all times.

6.5.7   Packaging of Fresh Produce

After harvest, the tissues of fruits and vegetables continue to undergo metabolic processes that affect their sensory attributes and ultimately, quality and shelf life. Respiration is the primary metabolic process that contributes to the postharvest physiological changes in the fruit or vegetable tissue. In this process, energy in the form of ATP is generated by the oxidation of glucose-rich compounds (e.g., starches) within the cells of the plants. Respiration in plants is typically an aerobic process, where oxygen is consumed and converted into carbon dioxide, water, and energy (in the form of heat). Plants can also undergo anaerobic respiration in certain situations, although this process tends to rapidly deteriorate the quality of the fruit or vegetable. In general, the respiration rate of a fruit or vegetable is proportional to the shelf life of the product, since this metabolic process is responsible for the senescence of the plant.

Many factors can influence the respiration rate of fruits and vegetables. In fruits, ethylene gas serves as a plant hormone that regulates the ripening and thus respiration process. Fruits are generally classified as either climacteric or non-climacteric. Climacteric fruits are fruits whose ripening is associated with an increase in both ethylene production and respiration, which results in an increase in carbon dioxide production. In climacteric fruits, ethylene acts as a trigger for the ripening process. Examples of climacteric fruits include apples, melons, bananas, avocadoes, and tomatoes. Non-climacteric fruits, on the other hand, do not increase ethylene production during the ripening process, and so, the presence of ethylene does not accelerate this process. Examples of non-climacteric fruits include citrus fruits and strawberries.

Many produce items are packaged using modified atmosphere packaging (MAP) technologies. MAP technologies generally fall under two separate categories: passive and active MAP technologies. In passive MAP, an atmospheric barrier is formed around a perishable fresh produce product. This barrier passively results in a modification of the interior atmosphere of the package. As the produce undergoes senescence, O_2 is consumed and CO_2 is released, and the barrier properties of the film allow the CO_2 to build thereby passively modifying the atmosphere, slowing down respiration and extending shelf life.

Active MAP is the utilization of scavenging materials (either within the packaging film or in the form of sachets) or the infusion of a specific composition of gasses into the package during sealing. These gaseous mixtures can be custom tailored to desired levels based on the respiration requirements of the produce item. Another form of active MAP involves one-way gas valves that allowed certain respiration gases to exit the package while preventing outside gases from entering the package. Roasted coffee beans are a great example of an active MAP package that utilizes a one-way valve. As roasted coffee beans age, they emit carbon dioxide. If carbon dioxide is allowed to build up to high levels within the package, staling will occur. This is why one-way gas valves are often incorporated into fresh coffee bean packages, to allow off gassing of carbon dioxide without preventing any influx of external gases.

Selection of passive or active MAP will depend upon how the fresh produce ripens and what quality attributes are impacted by that process versus the need to quickly halt ripening and respiration early in the shelf life. When oriented polypropylene of different thickness (20, 40, and 80 μm) was used to package ready-to-eat table grapes, it was determined that passive MAP of the highest thickness film attained 70 days of shelf when stored at 5 °C, which was more than active MAP was able achieve (Costa et al. 2011). Fresh endive stored at 20 °C under active MAP

Table 6.1 Examples of optimum MAP storage conditions for various fruits and their expected shelf lives

Fruit	Storage temperature (°C)	MAP conditions		Shelf life (Days)
		% CO_2	% O_2	
Apple	3	3	3	200–300
Avocado	7	5	10	12–56
Banana	15	2	5	21–60
Grapes	2	5	3	40–90
Lemon	15	5	5	130–220
Orange	10	10	5	42–84
Papaya	13	5	8	14–35
Pineapple	15	5	10	12–15
Strawberry	15	10	20	7–15

Adapted from Mangaraj and Goswami (2009)

created by the use of an O_2 scavenging sachet, did not change O_2 and CO_2 partial pressure during the steady-state period, compared to passive MAP, but induced a 50% reduction of the transient period and delayed greening and browning (Charles et al. 2008). Active MAP of romaine lettuce with a gas mixture of 10% O_2, 10% CO_2, and 80% N_2 delayed growth of autochthonous lettuce microflora, but not *Salmonella* and even favored the survival of the pathogen, possibly due to the elimination of its natural antagonists (Horev et al. 2012). The effects of the passive MAP on lettuce were less pronounced. These varying results highlight the importance of selecting MAP conditions that are specific to the respiration requirements of the produce item in question, while considering the intended outcome of the MAP package itself (Table 6.1).

6.5.8 Packaging of Dairy and Cheese Products

Fresh pasteurized milk, once bottled in glass, has been regularly packaged in blow molded polymers since the 1960s. The vast majority of fluid milk is packaged in HDPE containers, although some polyethylene terephthalate (PETE) and low-density polyethylene (LDPE) examples can be found. Half gallons are also sold in a three-layered carton format (polyethylene, paperboard, polyethylene) with a shelf life of 15–21 days at refrigeration temperatures (Fromm and Boor 2004).

Fermented dairy products such as block cheeses are often packaged in multi-layered co-polymer materials that are designed to exhibit a variety of properties to enhance the shelf life of these foods. Cheeses are often packaged in reduced oxygen environments in order to prevent unwanted growth of microbial spoilage organisms, such as mold. High carbon dioxide flushing of MAP packaged cheese is also common, as it displaces oxygen and assists in shelf life extension by inhibiting growth of mold and other spoilage organisms. In these conditions, CO_2 levels can range from 50% to 100%. The typical shelf life of high CO_2 MAP packaged hard cheeses (such as cheddar) is 4–6 weeks depending on storage conditions. High CO_2 conditions are also highly effective at extending the shelf life of cottage cheeses, which typically have a shelf life of 2–4 weeks. When cottage cheese packages are flushed with high CO_2, shelf life extended to approximately 8–12 weeks, depending on storage conditions. However, it has been reported that cottage cheese can take on a "soda-like" flavor, as the CO_2 absorbs into the product itself. Cheese packages also usually contain a moisture barrier to prevent the product from drying out. Nylon is becoming more popular in many co-polymeric cheese packages due to its enhanced durability and strength. This is very important for packaging of cheese blocks, as the sharp corners can puncture packages during transportation and handling.

6.6 Active Packaging Technologies

6.6.1 Overview

Packaging materials have evolved from providing a simple inert, passive containment for food products into packaging that can provide "active functions" such as extending shelf life and maintaining or improving desirable conditions of the packaged food. Active packaging is designed to deliberately release components into the packaged food or the surrounding environment. Two popular active packaging concepts include materials or packages that scavenge unwanted substrates or gases from the packaging environment, or materials or packages that emit gases or antimicrobials into the food or package environment (Suppakul et al. 2003). As mentioned previously in this chapter, the spoilage of many meat and food products is attributed to the production of unpleasant off odors. These off odors are generated either by bacterial growth and proliferation, endogenous enzyme activity, or by degradative chemical reactions (e.g., oxidation of lipids). Regardless of the mechanism, these off odors contribute to a reduction in product shelf life. Thus, it is evident that active packaging technologies such as odor-scavenging materials and antimicrobial packaging have the potential to positively influence and extend the shelf life of these products.

For an active packaging technology to be commercially viable, it must have several desirable attributes. First and foremost, the active components must be approved for use in direct food contact applications, meaning it will be non-toxic, non-allergenic, and harmless to human health. The active components must be easily incorporated into the package, by incorporation into the film, tray, sachet, label, and must be able to function under conditions of use for the package (e.g., shipment and storage temperatures, and lighting conditions). The active components also must not adversely impact the organoleptic profile of the product, nor mask any types of spoilage. Finally, the active components should be invisible to the customer. This is particularly a challenge for active components that are directly incorporated into a film, as this can cause a film become opaque, or otherwise degrade the optics of the film to such an extent it becomes unfit for packaging applications. This is especially the case where the film customer or final user desires a clear film in which the contents of the package can be visually inspected from outside the package. A few examples of active packaging technologies are discussed below.

6.6.2   Oxygen Scavenging Technologies

The shelf life of oxygen-sensitive food (e.g., fatty fish, fatty cuts of beef, and packaged nuts) can be extended by reducing the amount of oxygen within a package. Vacuum and MAP packaging technologies often extend shelf life based on this concept. Additionally, a variety of active packaging technologies have been developed that scavenge the available oxygen within a package. One approach is through the inclusion of a material (other than the package itself) capable of consuming oxygen. These materials usually consist of sachets with oxygen scavenging properties. Modern oxygen scavenging sachets usually contain a mixture of iron and sodium chloride. Moisture within a package activates the oxygen scavenging material by reacting with iron and oxygen to form iron oxide, also known as rust. This process is accelerated by the presence of sodium chloride, which acts as a catalyst for the rusting process. Using this process, the oxygen level within a package can be reduced to 0.01% or lower (Brandon et al. 2009). An advantage of this iron-based oxygen scavenging system is that it does not produce any offensive off odors or unwanted flavors. A disadvantage of this system is that a moisture content of greater than 50% is required for the sachets to work, rendering them inappropriate for use in packaged dry products (Brandon et al. 2009).

More recently, sachet free oxygen scavenging technologies have been introduced into the market. These sachet-free technologies rely on oxygen scavenging polymers within the film itself to extend product shelf life. A major advantage of this approach is the prevention of accidental ingestion of oxygen scavenging sachets. One patented approach to creating an oxygen scavenging film is to incorporate a resin with a large number of unsaturated bonds into the barrier layer. The film is then irradiated, which "triggers" the oxygen scavenging effect of the film by increasing the reactivity of the unsaturated bonds with ground state oxygen (Beckwith et al. 2016). A frequent application of oxygen scavenging films is in packaging of high-value bread, cake, and cereal products, especially those using a preservative-free formulation. In these products, oxygen scavenging films can extend shelf life by preventing the growth of several fungal spoilage organisms, including *Penicillium* spp. (Nielsen and Rios 2000). Additional examples of oxygen scavenging technologies are provided below in Table 6.2.

Table 6.2 Select examples of commercially available oxygen scavenging technologies

Oxygen scavenging technology	Delivery method	Commercial product examples
Metal powders (Fe, cu, etc.)	Sachet, labels, tray, extrusion films	Ageless® series—Mitsubishi gas and chemical company Sorbent system—Impak corporation O-busters®—Sachets and strips - Dessicare Inc. O₂Block®—Nanobiomatters Oxy-guard®—Clariant FreshMax® labels and sachets—Multisorb technologies ShelfPlus O₂®—Albis
Ascorbic acid and ascorbate salts	Sachets and gaskets	Celox® series—GCP applied technologies Darex® MB2003—GCP applied technologies
Catechol	Sachet	Tamotsu—OhE chemicals, Inc.
Photosensitive dyes	Film	Zero₂—CSIRO Australia
Enzymes, e.g., glucose oxidase, alcohol oxidase	Sachet	Bioka—Bioka Ltd.
MXD-6 (nylon)	Film, bottles	Oxbar®—Plastipack
Unsaturated polymers	Extrusion films, trays, bottles	Amosorb®—PolyOne Cryovac® OS films—Sealed Air Corporation
Pd catalysts/H₂	Sachets, films, labels	HyGuard™—Polyone

6.6.3 Carbon Dioxide Scavenging Technologies

While CO_2 is often incorporated into modified atmosphere packages for its bacteriostatic properties, sometimes the buildup of this gas within a package is a highly undesirable property. In some foods, for example, cottage cheeses, buildup of CO_2 can have a negative impact on some of the flavor attributes of the product which negatively impact shelf life. In other foods, especially fermented foods, the buildup of CO_2 over time can compromise package integrity. In each case, CO_2 scavenging technologies can serve as a solution to the problem.

Broadly speaking, CO_2 scavenging technologies fall into one of two categories by mode of action: physical and chemical absorption. In chemical absorption, calcium salts and alkaline solutions can be utilized to react with residual CO_2 in a package. The most frequently used salt for this reaction is calcium hydroxide, which has favorable properties that make it suitable for food contact use. Sodium carbonate is also used, although this compound is only suitable for scavenging CO_2 from moist package environments, since water is a requirement for the chemical reaction responsible for the consumption of CO_2. In physical absorption, CO_2 gas molecules adsorb onto a physical substrate and become trapped in the physical network of the substrate. The two most common substrates used for this purpose are activated carbon and zeolites.

Activated carbon is characterized by an amorphous porous structure with an extremely large surface area. An advantage of activated carbon is that its non-polar structure means that moisture inside a package has relatively little impact on the rate of CO_2 adsorption. Zeolites, which are microporous, aluminosilicate minerals with a complex three-dimensional structure, are also used for physical adsorption of CO_2 gas. The surface area and structure of zeolites can be customized by varying the cation (e.g., sodium and calcium) used for formation of the zeolite. Many zeolites have a higher affinity for water vapor than CO_2 gas, which can be a disadvantage for the use of zeolites as CO_2 scavengers in high moisture food products.

There are numerous examples of the successful use of CO_2 scavenging technologies to extend shelf life of foods. Some fruits, such as pears, can undergo accelerated browning reactions when CO_2 levels build up to high levels within a package. In one study, researchers found pears packaged in CO_2 scavenging film and subjected to long-term cold storage had enhanced shelf life versus pears store in non-scavenging film (Nugraha et al. 2015).

6.6.4 Odor Scavenging Packaging

Odor scavenging technologies are designed specifically to control undesirable odors within a package (de Abreu et al. 2011). Plastic polymer-based packaging has long been known to absorb volatile compounds from food, which is commonly referred to as "scalping" (Van Willige et al. 2002). Conversely, plastic packaging materials can also impart undesirable volatile compounds or odors into food. Although generally regarded as a negative attribute, it is possible to construct a package where the absorption of compounds by food becomes a desired feature in controlling odor profiles. For some foods, the oxidation of lipids can result in the development of "rancid" odors. These products, such as nuts and oils, benefit from using a combination of oxygen scavengers and odor absorbers in their packaging. Other foods such as bone-in-meats, sausage, ham, salami, pepperoni, poultry, and processed poultry meats such as turkey pepperoni can generate sulfur type off odors during distribution and storage. Hydrogen sulfide and other sulfur containing compounds, such as thiols or mercaptans, are generated during the normal shelf life of these products as the by-products of enzymatic or microbial degradation of sulfur containing amino acids such as cysteine. These odors are extremely unpleasant, especially hydrogen sulfide (rotten egg) and ethyl mercaptan (skunk). These off odors are particularly a problem in high oxygen barrier packaging, which effectively traps the off odor inside the package. Although the product may still be perfectly safe for consumption, the odors are released upon opening the package, causing consumers to regard the product as spoiled.

Other off odors may be composed of additional chemical classes, such as aldehydes, acids, ketones, and amines. For example, fish and seafood generate amine odors while dairy products may generate sour or "buttery" odors. Thus, it becomes important to know the profile for undesirable volatile compounds for the products being packaged, so that an appropriate scavenger system can be used. Two important caveats to designing an active odor-scavenging package are that the scavenger materials not "scalp" the desirable odors from the packaged product and that the use of the scavenger material does not mask product spoilage.

Being volatile compounds, odors may be removed within a packaging system by physical means through adsorption or absorption. Adsorption is a physical and/or chemical process by which one substance becomes attached to another through physical and/or chemical interactions. More specifically, adsorption is the adherence, binding, or attraction of atoms, molecules, or ions to the surface of another material. In adsorption, binding to the surface is frequently reversible, but compounds that have taste or odor typically tend to bind strongly. Adsorbents are frequently characterized by having very large surface areas per unit weight. Typical adsorbents include activated carbon and silica. Absorption is the penetration of one substance into the inner structure of another. The most common industrial absorbents include zeolites, molecular sieves, and cyclodextrins.

Odor compounds may also be removed within a package using chemical means. This is known as chemisorption. During a spontaneous reaction, molecular bonds may be broken or created as the volatile odor compounds react with the scavenging agent. These processes could include synthesis reactions, reduction/oxidation reactions, or acid–base reactions. For example, one method for the removal of hydrogen sulfide is through a reaction with a metal ion such as iron or copper, resulting in the formation of a metal sulfide. An overview of typical odor scavenging technologies and their applications is listed in Table 6.3.

Table 6.3 Frequently used odor scavenging applications

Scavenging agent	Mechanism(s)	Application
Activated carbon	Adsorption	Sachet, pad, label, tray
Alumina	Adsorption Chemisorption	Sachet, pad, label, tray, film
Baking soda	Chemisorption	Sachet, pad, label
Cyclodextrins	Absorption	Sachet, pad, label, tray, film
Layered double hydroxides (LDH), e.g., hydrotalcite	Adsorption, absorption, chemisorption	Sachet, pad, label, tray, film
Metals: e.g., cu, Zn, Fe, Ni, Ag	Chemisorption	Sachet, pad, label, tray, film
Metal oxides/salts: e.g., ZnO, MgO CaO, Ca(OH)$_2$, Fe$_2$O$_3$	Chemisorption	Sachet, pad, label, tray, film
Molecular sieves	Absorption Adsorption	Sachet, pad, label, tray, film
Organic acids, e.g., citric acid, ascorbic acid	Chemisorption	Sachet, pad, label
Polyalkylene imine, e.g., polyethylene imine	Chemisorption	Film, tray, label
Silica gel	Absorption Adsorption	Sachet, pad, label, tray, film
Zeolites	Absorption	Sachet, pad, label, tray, film
Zinc Ricinoleate	Chemisorption	Sachet, pad, label, tray, film

6.6.5 Moisture Control Technologies

For many foods, excess moisture within a package can have a major negative impact on shelf life. In dry powdered products, excess moisture can cause clumping and caking to occur and in hard candies, and can soften the product, leading to a highly undesirable mouth feel. In many packaged meat products, moisture loss in the form of purge occurs over time and must be controlled. Ultimately, moisture control technologies help control the water activity within a package, which minimizes both microbial growth and unwanted sensory changes within a product. Desiccants are often incorporated into packages of dry food products as sachets. Silica gels, calcium oxide, and clays are frequently used as desiccants. Absorbent pads are frequently used in fresh meat packages to absorb excess purge. These are typically made of silica gel or cellulosic fibers.

6.6.6 Antimicrobial Packaging Technologies

Antimicrobial packaging technologies extend the shelf life of food by killing unwanted spoilage organisms that contribute to loss of sensory attributes. Additionally, antimicrobial packaging technologies may be designed to kill or help control growth of pathogens, such as *Listeria monocytogenes, Salmonella* spp. and *Escherichia coli,* which may be present on the surface or within food. Antimicrobial coatings and films are some of the more promising antimicrobial packaging technologies and have been researched extensively for several decades. In this approach, the antimicrobial ingredient is incorporated directly into the coating or film, which is then in direct contact with the food. Table 6.4 below shows several recent research examples of antimicrobial coatings and films and their applications in food packaging.

When the antimicrobial film is in direct contact with a food surface, the antimicrobial ingredient diffuses into the food which prevents microbial growth. Because of this, antimicrobial films must be in direct contact with a food for antimicrobial activity to take place. A key challenge with designing these antimicrobial films involves calibrating the rate of diffusion of the antimicrobial ingredient into the food. If the rate of diffusion is too fast, then the antimicrobial activity is quickly depleted. If the rate of diffusion is too slow, minimal antimicrobial activity may occur. Another challenge is that many food matrices may inactivate or reduce the antimicrobial activity of these active ingredients. For example, fatty foods may reduce the antimicrobial activity of many bacteriocins, such as nisin, which is effective against Gram-positive bacteria, such as *L. monocytogenes* (Franklin et al. 2004).

Many approaches to incorporating antimicrobial ingredients into films and coatings have been explored. Edible films, typically made of cellulosic materials, have been formulated with naturally derived bacteriocins, such as nisin. Nisin has extremely potent antimicrobial activity against Gram-positive bacteria, such as *L. monocytogenes*. When individually packaged hot dogs were coated with cellulose materials con-

Table 6.4 Select references on antimicrobial agents incorporated into films and coatings

Antimicrobial ingredient	Food contact matrix	Foods	Targeted microorganisms	References
Oregano and thyme essential oils	LDPE	Culture only	*Salmonella* Typhimurium, *Escherichia coli, Listeria monocytogenes*	Solano and de Rojas Gante (2012)
Nisin	Cellulose coating	Frankfurters	*Listeria monocytogenes*	Franklin et al. (2004)
Lauroyl arginate ethyl (LAE)	Chitosan and gelatin	Culture only	*Salmonella* Typhimurium, *Escherichia coli, Listeria monocytogenes, Campylobacter jejuni*	Haghighi et al. (2019)
Rosemary and cinnamon	Whey protein-based film	Culture only	*E. coli, S. aureus,* and *Penicillium* spp.	Ribeiro-Santos et al. (2017)
Bacteriocin 729	Poly-lactic acid film	Fish filets	*Listeria monocytogenes, Staphylococcus aureus, Pseudomonas aeruginosa, Aeromonas hydrophila, Escherichia coli, Salmonella* Typhimurium	Woraprayote et al. (2018)

taining nisin, a 2-$\log_{10}$ reduction in *Listeria monocytogenes* counts was observed over a 60-day storage window (Franklin et al. 2004). Researchers have also incorporated nisin into PVC and linear low-density polyethylene (LLDPE) into packaging systems, and observed that it significantly reduced populations of *Salmonella* Typhimurium when used to package fresh poultry (Natrajan and Sheldon 2000). While nisin has shown good antimicrobial activity across a broad spectrum of bacteria, one concern involves the risk of bacteria becoming resistant to its activity via mutations. To address this concern, antimicrobial film approaches using nisin often include a second bacteriocin in order to reduce the risk of resistance causing mutations from occurring.

Essential oils such as those derived from oregano, thyme, cinnamon, and pimento, have long been known to exhibit antimicrobial properties, and are often incorporated into films in a research setting. In one study, a 4.0% (w/w) combination of oregano and thyme oils incorporated into extruded LLDPE films was able to inhibit *Salmonella Typhimurium, Listeria monocytogenes,* and *Escherichia coli* O157:H7 using standard culture methods (Solano and de Rojas Gante 2012). One major disadvantage of essential oils comes from their impact on the sensory attributes of a food product. While these compounds are highly active against a wide range of bacteria, they tend to confer major sensory properties (flavor, aroma) into a food.

While research on antimicrobial films for spoilage organisms and pathogen control has been promising, few technologies and approaches have successfully transitioned into the real world. One issue is that laboratory evidence rarely translates into real-world efficacy. For example, many laboratory experiments are reliant upon food simulants, rather than complex food matrices. In real-world food systems, the composition of pH, salt, carbohydrate, protein, and fats within a food can drastically change, which may impact the performance of the antimicrobial film. Storage conditions of the packaged food itself can also have a major impact on the performance of the antimicrobial films. This is mainly due to temperature and humidity having a major impact on the rate of diffusion of antimicrobial ingredients. As diffusion slows, so does the antimicrobial activity. When promising antimicrobial films are adapted from the laboratory into a real-world setting, what is often seen is an initial spike in antimicrobial efficacy, which quickly falters due to lack of diffusion of the antimicrobial into the food, a negative interaction of the antimicrobial with food components, or a saturation of the food with the antimicrobial ingredient.

Finally, a major challenge with antimicrobial films regards their difficulty in manufacturing. Most laboratory-scale experiments are performed using a casting process, rather than extrusion. Extrusion is the production process of choice for polymeric films due to its high line speeds and ability to create a uniform product. The high heat and pressures encountered during extrusion often destroy any antimicrobial activity of many antimicrobial ingredients. Future research investigating novel compounds should always incorporate stability testing under various temperatures and pressures to ensure compatibility with the extrusion process.

6.7   Intelligent Packaging

According to the United States Department of Agriculture, each year in the United States, about 30–40% of the entire food supply is wasted (USDA 2019). This represents 133 billion pounds of food and is valued at over $162 billion. An emerging area of packaging, termed "intelligent packaging," seeks to use innovative technologies to help reduce food waste.

Intelligent packaging has been defined as "materials and articles that monitor the condition of packaged food or the environment surrounding the food" (EFSA 2009). In other words, intelligent packaging technologies seek to communicate information about a food package to the consumer so that they may make an informed decision. In general, intelligent packaging falls under three broad categories: (1) data storage devices, (2) indicators, and (3) sensors.

Intelligent packaging technologies that monitor data are intended to improve supply chain efficiency. The most frequently encountered data monitoring approaches are barcodes and Radio Frequency Identification Devices (RFID) tags. Barcodes utilize patterns to store information about a package. They are widely used to assist in inventory control and have also experienced increasing adoption for traceability purposes. RFID tags are more expensive but are able to store far more data than barcodes. RFID tags are becoming increasingly adopted for use in inventory management, and for the promotion of safe food by enhancing traceability in the event of a foodborne illness outbreak. RFID tags consist of a small antenna wired to a small microprocessor, allowing the tag to communicate to other devices via radio waves.

Indicators communicate the presence or absence of an analyte or a condition to the consumer. Often, they do this via change in color. Indicators may be placed on or outside a package. Time and Temperature indicators (TTIs) and freshness indicators are two examples of commercially available intelligent packaging devices. TTIs monitor the storage of a product along the production chain, specifically, the temperature and time of holding throughout a process, usually transportation. As discussed previously, time and temperature are two critical variables that have a negative impact on the shelf life of a product if they are not controlled. TTIs help to keep transportation processes in control with respect to time and temperature, thus leading to food that retains its quality better. TTIs can monitor a critical temperature or they can monitor the history of time and temperature along an entire process.

Freshness indicators communicate information about the quality of a food to the consumer. These devices often monitor microbiological metabolites that are correlated with a loss of quality. For example, volatile nitrogen compounds and biogenic amines are often associated with the loss of quality of a meat product. When the amount of metabolite exceeds a threshold, a chemical reaction takes place that leads to a color change of the freshness indicator. An example of this is the SensorQ sensor from Food Quality Sensor International Inc., (Lexington, MA). This sensor is placed inside of a package and monitors the level of amines generated by microbiological degradation of a meat product. Amine oxidases present in the sensor react to the presence of amines by initiating a change of color, communicating to the consumer that the product is no longer fresh.

Sensors are different from indicators, in that they use energy to trigger a response. They are designed to quantify or detect key analytes of interest. A sensor receives a signal on its receptor from the analyte. This signal is then converted into energy that controls the communicative ability of the device. Chemical sensors receive a signal from a chemical analyte, for example, the amount of CO_2 within a package. Biosensors receive the signal from a biological material, such as the presence of an antigen that may be associated with a pathogen.

6.8 Concluding Thoughts

Packaging has grown to play a major role in everyday consumer life. Significant advances in food packaging technology over the last century have assisted in enhancing the shelf life of many products, all while providing an elevated level of convenience to the consumer. The future should focus on adapting innovative technologies that may control pathogenic and spoilage organisms to the real world. As we have seen, many of these technologies are promising in laboratory settings, but they do not always behave the same when translated into real-world applications. Additionally, while current packaging technologies assist in the enhancement of shelf life, the fact that food waste is still a major global concern suggests that research and innovation in this area need to continue. Future research should also focus on sustainability and improving the environmental impact of packaging materials. Unfortunately, many of the materials frequently used in food packaging are not recyclable, which can have a detrimental impact on the environment. Though sometimes the benefits of food packaging go unnoticed compared to the above limitations, the importance of these materials in our everyday lives cannot be understated.

References

Ahmed, Z., O.N. Donkor, W.A. Street, and T. Vasiljevic. 2013. Activity of endogenous muscle proteases from 4 Australian underutilized fish species as affected by ionic strength, pH, and temperature. *Journal of Food Science* 78 (12): C1858–C1864. https://doi.org/10.1111/1750-3841.12303.

Beckwith, S., F. B. Edwards, J. Rivett, C. L. Ebner, T. Kennedy, , R. McDowell, and D. V. Speer. 2016. Multilayer film having an active oxygen barrier layer with radiation enhanced active barrier properties. United States Patent 9452592. Cryovac, Inc. Duncan, SC

Boziaris, I.S., and F.F. Parlapani. 2017. Specific spoilage organisms (SSOs) in fish. In *The Microbiological Quality of Food Foodborne Spoilers*, ed. Antonio Bevilacqua, Maria Rosaria Corbo, and Milena Sinigaglia, 61–98. Amsterdam: Elsevier. https://doi.org/10.1016/B978-0-08-100502-6.00006-6.

Brandon, K., M. Beggan, P. Allen, and F. Butler. 2009. The performance of several oxygen scavengers in varying oxygen environments at refrigerated temperatures: Implications for low-oxygen modified atmosphere packaging of meat. *International Journal of Food Science & Technology* 44 (1): 188–196. https://doi.org/10.1111/j.1365-2621.2008.01727.x.

Calafat, A.M., Z. Kuklenyik, J.A. Reidy, S.P. Caudill, J. Ekong, L.L. Needham, and L. L. 2005. Urinary concentrations of Bisphenol A and 4-Nonylphenol in a human reference population. *Environmental Health Perspectives* 113 (4): 391–395. https://doi.org/10.1289/ehp.7534.

Charles, F., C. Guillaume, and N. Gontard. 2008. Effect of passive and active modified atmosphere packaging on quality changes of fresh endives. *Postharvest Biology and Technology* 48 (1): 22–29.

Costa, C., A. Lucera, A. Conte, M. Mastromatteo, B. Speranza, A. Antonacci, and M.A. Del Nobile. 2011. Effects of passive and active modified atmosphere packaging conditions on ready-to-eat table grape. *Journal of Food Engineering* 102 (2): 115–121.

de Abreu, D.A.P., J.M. Cruz, and P.P. Losada. 2011. Active and intelligent packaging for the Food industry. *Food Reviews International* 28 (2): 146–187. https://doi.org/10.1080/87559129.2011.595022.

De Poix, H.M.J.T. 1945. *Process for Preserving Perishable Foodstuffs*. Dewey and Almy Chemical Company: Cambridge, MA.

Del Nobile, M.A., and A. Conte. 2013. *Packaging for Food Preservation*. New York, NY: Springer. https://doi.org/10.1007/978-1-4614-7684-9.

DeWitt, C., and A. Oliveira. 2016. Modified atmosphere systems and shelf life extension of fish and fishery products. *Food* 5 (4): 48. https://doi.org/10.3390/foods5030048.

Djenane, D., and P. Roncalés. 2018. Carbon monoxide in meat and fish packaging: Advantages and limits. *Food* 7 (2): 12. https://doi.org/10.3390/foods7020012.

European Food Safety Authority (EFSA). 2009. Guidelines on submission of a dossier for safety evaluation by the EFSA of active or intelligent substances present in active and intelligent materials and articles intended to come into contact with food. *EFSA Journal* 7 (8): 1208.

Farber, J.M. 2016. Microbiological aspects of modified-atmosphere packaging technology - A review. *Journal of Food Protection* 54 (1): 58–70. https://doi.org/10.4315/0362-028X-54.1.58.

Font-i-Furnols, M., and L. Guerrero. 2014. Consumer preference, behavior and perception about meat and meat products: An overview. *Meat Science* 98 (3): 361–371. https://doi.org/10.1016/j.meatsci.2014.06.025.

Franklin, N., K.D. Cooksey, and K. Getty. 2004. Inhibition of *Listeria monocytogenes* on the surface of individually packaged hot dogs with a packaging film coating containing Nisin. *Journal of Food Protection* 67 (3): 480–485.

Fromm, H.I., and K.J. Boor. 2004. Characterization of pasteurized fluid milk shelf-life attributes. *Journal of Food Science* 69 (8): M207–M214.

Gram, L. 2001. Potential hazards in cold-smoked fish: *Clostridium botulinum* type E. *Journal of Food Science* 66 (35): S1082–S1087. https://doi.org/10.1111/j.1365-2621.2001.tb15527.x.

Haghighi, H., R. De Leo, E. Bedin, F. Pfeifer, H.W. Siesler, and A. Pulvirenti. 2019. Comparative analysis of blend and bilayer films based on chitosan and gelatin enriched with LAE (lauroyl arginate ethyl) with antimicrobial activity for food packaging applications. *Food Packaging and Shelf Life* 19: 31–39.

Heyerick, A., Y. Zhao, P. Sandra, K. Huvaere, F. Roelens, and D. De Keukeleire. 2003. Photolysis of hop-derived trans-iso-alpha-acids and trans-tetrahydroiso-alpha-acids: Product identification in relation to the lightstruck flavour of beer. *Photochemical and Photobiological Sciences* 2 (3): 306–314.

Horev, B., S. Sela, Y. Vinokur, E. Gorbatsevich, R. Pinto, and V. Rodov. 2012. The effects of active and passive modified atmosphere packaging on the survival of *Salmonella enterica* serotype Typhimurium on washed romaine lettuce leaves. *Food Research International* 45 (2): 1129–1132.

Kelly, C.A., M. Cruz-Romero, J.P. Kerry, and D.P. Papkovsky. 2018. Assessment of performance of the industrial process of bulk vacuum packaging of raw meat with nondestructive optical oxygen sensing systems. *Sensors* 18 (5): 1395. https://doi.org/10.3390/s18051395.

Lucquin, A., K. Gibbs, J. Uchiyama, H. Saul, M. Ajimoto, Y. Eley, A. Radini, C.P. Heron, S. Shoda, Y. Nishida, J. Lundy, P. Jordan, S. Isaksson, and O.E. Craig. 2016. Ancient lipids document continuity in the use of early hunter-gatherer pottery through 9,000 years of Japanese prehistory. *Proceedings of the National Academy of Sciences of the United States of America* 113 (15): 3991–3996. https://doi.org/10.1073/pnas.1522908113.

Mangaraj, S., and T.K. Goswami. 2009. Modified atmosphere packaging of fruits and vegetables for extending shelf-life: A review. *Fresh Produce* 3 (1): 1–31.

Natrajan, N., and B.W. Sheldon. 2000. Efficacy of nisin-coated polymer films to inactivate *Salmonella* Typhimurium on fresh broiler skin. *Journal of Food Protection* 63 (9): 1189–1196.

Nielsen, P.V., and R. Rios. 2000. Inhibition of fungal growth on bread by volatile components from spices and herbs, and the possible application in active packaging, with special emphasis on mustard essential oil. *International Journal of Food Microbiology* 60 (2–3): 219–229. https://doi.org/10.1016/S0168-1605(00)00343-3.

Nugraha, B., N. Bintoro, and H. Murayama. 2015. Influence of CO_2 and C_2H_4 adsorbents to the symptoms of internal browning on the packaged 'silver bell' pear (*Pyrus communis* L.). *Agriculture and Agricultural Science Procedia* 3 (1): 127–131.

Oulé, M.K., K. Tano, A.-M. Bernier, and J. Arul. 2006. *Escherichia coli* inactivation mechanism by pressurized CO_2. *Canadian Journal of Microbiology* 52 (12): 1208–1217. https://doi.org/10.1139/w06-078.

Ribeiro-Santos, R., A. Sanches-Silva, J.F.G. Motta, M. Andrade, I. de Araújo Neves, R.F. Teófilo, M.G. de Carvalho, and N.R. de Melo. 2017. Combined use of essential oils applied to protein base active food packaging: Study in vitro and in a food simulant. *European Polymer Journal* 93: 75–86.

Risvik, E. 2001. The Food and I sensory perception as revealed by multivariate methods. In *Food, people and society*, ed. L.J. Frewer, E. Risvik, and H. Schifferstein, 23–37. Berlin, Heidelberg: Springer. https://doi.org/10.1007/978-3-662-04601-2_3.

Robertson, G.L. 2012. *Food Packaging: Principles and Practice*. 3rd ed. Boca Raton, FL: CRC Press Taylor & Francis.

Rossaint, S., S. Klausmann, U. Herbert, and J. Kreyenschmidt. 2014. Effect of package perforation on the spoilage process of poultry stored under different modified atmospheres. *Food Packaging and Shelf Life* 1 (1): 68–76. https://doi.org/10.1016/j.fpsl.2014.01.002.

Russell, S.M., D.L. Fletcher, N.A. Cox, and N. A. 1995. Spoilage Bacteria of fresh broiler chicken carcasses. *Poultry Science* 74 (12): 2041–2047. https://doi.org/10.3382/ps.0742041.

Schug, T. T., J. J. Heindel, L. Camacho, K. B. Delclos, , P. Howard, A. F. Johnson,, J. Aungst, D. Keefe, R. Newbold, N. J. Walker, R. Thomas Zoeller, and J. R. Buchen. 2013. A new approach to synergize academic and guideline-compliant research: The CLARITY-BPA research program. Reproductive Toxicology 40:35–40. doi:https://doi.org/10.1016/j.reprotox.2013.05.010.

Seachrist, D.D., K.W. Bonk, S.-M. Ho, G.S. Prins, A.M. Soto, and R.A. Keri. 2016. A review of the carcinogenic potential of bisphenol A. *Reproductive Toxicology* 59: 167–182. https://doi.org/10.1016/j.reprotox.2015.09.006.

Solano, A.C.V., and C. de Rojas Gante. 2012. Two different processes to obtain antimicrobial packaging containing natural oils. *Food and Bioprocess Technology* 5 (6): 2522–2528.

Strom, E.T., and S.C. Rasmussen. 2011. *100+ Years Of Plastics. Leo Baekeland and Beyond.* Washington, DC: American Chemical Society. https://doi.org/10.1021/bk-2011-1080.

Suppakul, P., J. Miltz, K. Sonneveld, and S.W. Bigger. 2003. Active packaging technologies with an emphasis on antimicrobial packaging and its applications. *Journal of Food Science* 68 (2): 408–420. https://doi.org/10.1111/j.1365-2621.2003.tb05687.x.

Thompson, A.K. 2010. *Controlled Atmosphere Storage of Fruits and Vegetables.* 2nd ed. Wallingford, Oxfordshire, UK: CABI.

United States Department of Agriculture. 2019. Estimates of Food Waste. https://www.usda.gov/foodwaste/faqs. Accessed 31 November 2019.

Van Willige, R., D. Schoolmeester, A. Van Ooij, J. Linssen, and A. Voragen. 2002. Influence of storage time and temperature on absorption of flavor compounds from solutions by plastic packaging materials. *Journal of Food Science* 67 (6): 2023–2031. https://doi.org/10.1111/j.1365-2621.2002.tb09495.x.

Voges, K.L., C.L. Mason, J.C. Brooks, R.J. Delmore, D.B. Griffin, D.S. Hale, W.R. Henning, D.D. Johnson, C.L. Lorenzen, R.J. Maddock, R.K. Miller, J.B. Morgan, B.E. Baird, B.L. Gwartney, and J.W. Savell. 2007. National beef tenderness survey–2006: Assessment of Warner–Bratzler shear and sensory panel ratings for beef from US retail and foodservice establishments. *Meat Science* 77 (3): 357–364. https://doi.org/10.1016/j.meatsci.2007.03.024.

Woraprayote, W., L. Pumpuang, A. Tosukhowong, T. Zendo, K. Sonomoto, S. Benjakul, and W. Visessanguan. 2018. Antimicrobial biodegradable food packaging impregnated with Bacteriocin 7293 for control of pathogenic bacteria in pangasius fish fillets. *LWT* 89: 427–433.

Zhao, C., B. Ge, J. De Villena, R. Sudler, E. Yeh, S. Zhao, D.G. White, D. Wagner, and J. Meng. 2001. Prevalence of *Campylobacter* spp., *Escherichia coli*, and *Salmonella* serovars in retail chicken, Turkey, pork, and beef from the greater Washington, D.C., area. *Applied and Environmental Microbiology* 67 (12): 5431–5436. https://doi.org/10.1128/AEM.67.12.5431-5436.2001.

Chapter 7
Beyond the Standard Plate Count: Genomic Views into Microbial Food Ecology

Sarah M. Hertrich and Brendan A. Niemira

7.1 Introduction

Genomic approaches to preserving and extending shelf life, as well as improving functional properties and nutritional quality.

For more than half a century, spoilage-causing microbes have been identified with some success primarily relying on traditional selective and nonselective culture techniques (Waite et al. 2009). Advancements in genomics research have helped scientists to better understand how microbes interact with one another and their environment, how they infect their host, and how they have evolved over time. In the field of food science, genomics have helped us to better understand the microbial ecology of microbial communities on foods and food contact surfaces. Microbial ecology is of special interest to food microbiologists because it describes the ability of microbes to function in complex food environments (Floros et al. 2010). While studying individual species is important, traditional cultivation methods of spoilage microbes can be laborious, time-consuming, and biased by selective germination and outgrowth conditions on nutrient agar plates (Davey 2011; Kort et al. 2008).

Microbial spoilage is a complex ecological process involving the presence of multiple species with specific niches, as well as metabolic dependencies, which can be easily overlooked using basic culture-dependent techniques which favor the growth of single microbial species (Gram et al. 2002). Understanding how these individual species function within a diverse microbial environment can help us to understand the "bigger picture" of how these species interact in complex biological systems. These dynamic microbial communities include nonpathogenic spoilage

S. M. Hertrich · B. A. Niemira (✉)
Food Safety Intervention Technologies Research Unit, Eastern Regional Research Center, USDA-ARS, Wyndmoor, PA, USA
e-mail: sarmark@udel.edu; Brendan.Niemira@USDA.GOV

© Springer Nature Switzerland AG 2021
P. J. Taormina, M. D. Hardin (eds.), *Food Safety and Quality-Based Shelf Life of Perishable Foods*, Food Microbiology and Food Safety,
https://doi.org/10.1007/978-3-030-54375-4_7

microbes which affect food qualities including taste, flavor, texture, appearance, and shelf life (Floros et al. 2010). Genomics can also help scientists to better understand the process of fermentation, which is the microbial degradation of organic compounds within food products and is one of the oldest and most commonly used forms of food preservation (Sieuwerts et al. 2008). Because food fermentations are typically carried out by mixed cultures consisting of multiple microbial species, understanding the population dynamics of these cultures can help to increase their industrial performance (Sieuwerts et al. 2008).

Culture-independent genomics tools, such as genome sequencing, have allowed scientists to more accurately estimate the microbial diversity of foods (Floros et al. 2010). Until recently, scientists were limited to looking at DNA sequences of a small set of genes among small set of organisms at one given time. Now, scientists are able to obtain complete information about the organization and genetic composition of entire genomes (Watson et al. 2008) at a considerably low cost. In addition to microbial genomics, which is the study of the genomes of individual microbes, microbial metagenomics is the culture-independent functional and sequence-based analysis of the heterogeneous microbial genomes from a particular environment (Riesenfeld et al. 2004; Handelsman et al. 1998; Casas and Rohwer 2007; Gilbert and Dupont 2011; Casas and Maloy 2014).

Other "omics" tools including proteomics and metabolomics provide a better understanding of metabolites, enzyme activity, and metabolic fluxes (Floros et al. 2010), which also play a role in food degradation. Changes to the environment, including deforestation, irrigation, coastal zone degradation, wetland modification, expansion of urban areas, global climate change, and other human interventions, can lead to changes in the structure, composition, and dynamics of microbial communities (Patz et al. 2004) in which our food is grown or produced. When used appropriately, "omics" tools can be used to help scientists generate models that can identify changes in microbial ecology and, therefore, prevent spoilage of foods and food crops (Casas and Maloy 2014).

7.2   Recombinant DNA Technology

Recombinant DNA (rDNA) technology is a nontraditional alternative to conventional breeding practices, which allows for the transfer of a large number of specific genes into edible plants (Kumar and Kumar 2015). This technology, also referred to as genetic engineering or transgenic technology, allows for the genetic modification of edible plants to prevent spoilage, increase crop yields, decrease exposure to herbicides and pesticides, decrease production costs, and ultimately the creation of more nutritious food products at cheaper prices. In addition, rDNA technology has also allowed for the reduction or elimination of allergens from foods in order to protect those individuals who suffer from food allergies (IFT 2000). It is believed that the use of rDNA technology will allow us to feed the world in the future, when the population will have multiplied considerably, and there will be less land and resources available to grow crops. Crops that are tolerant to local soil and weather conditions, particularly in developing countries, will allow for the growth of nutritious foods at an affordable cost.

Recombinant DNA technology involves the "cloning" of a specific genomic target into a "transformed" organism of interest. Cloning of bacterial DNA describes the ability to construct recombinant DNA molecules and maintain them within cells. This process typically involves the use of a vector which provides all the necessary information that allows the new DNA to be inserted into the cell, allows the recombinant vector to multiply independently, and carries a selectable marker which distinguishes transformed cells from non-transformed cells (Watson et al. 2008). The most common types of vectors are plasmids, which are small (~3 kb) circular DNA molecules found naturally in many bacteria. Restriction enzymes are used to cut specific segments of the vector DNA where the recombinant DNA is to be inserted and ligation enzymes join the ends of the inserted DNA sequence to the new cell.

7.3   Whole Genome Sequencing

In 1953, Watson and Crick solved the structure of DNA by studying the crystallographic data produced by Rosalind Franklin and Maurice Wilkins (Watson and Crick 1953; Zallen 2003). This discovery allowed future scientists to understand the concepts of DNA replication and encoding proteins in nucleic acids; however, the ability to "read" DNA sequences did not develop for quite some time (Heather and Chain 2016). The first studies of microbial genomes began in the 1970s with the sequencing of two bacteriophage genomes (Wooley et al. 2010; Fiers et al. 1976; Sanger et al. 1978). The development that was considered the "major breakthrough" in DNA sequencing technology came about in 1977, known as Sanger's "chain termination" or dideoxy technique (Sanger and Nicklen 1977). Then, in 1995 the first bacterial genome, *Haemophilus influenza*, was sequenced and regarded as a huge step for the microbiology community (Wooley et al. 2010; Fleischmann et al. 1995).

Advanced molecular techniques, specifically whole genome sequencing, have helped scientists to better understand gene expression in individual microbial cells (Floros et al. 2010). Whole genome sequencing arrays have also allowed scientists to observe the complex regulation of gene expression within the cell that occurs during infection (Toledo-Arana et al. 2009). For example, scientists now understand exactly which virulence genes in specific species of bacteria and viruses are upregulated in the host in order for infection to occur.

With information provided by whole genome sequencing, food scientists will be able to develop cost-effective approaches for detection and control of the strains that are likely to cause disease (Floros et al. 2010). Sequencing of plant and animal genomes will allow scientists to study their evolution and domestication and, therefore, facilitate establishment of more effective breeding programs (Qin et al. 2014). Specialized intervention technologies in food processing plants, as well as other points in the food production chain, can also be developed to prevent contamination with foodborne pathogens (Floros et al. 2010). Strain monitoring may help identify dominant microbial strains which drive food spoilage, as well as strains that drive the fermentation process (Ecrolini 2013), which would benefit future food production and sustainability strategies. For example, scientists used whole genome sequencing coupled with traditional culturing techniques to identify the specific spoilage organisms in vacuum packaged ham (Piotrowska-Cyplik et al. 2017). They were also able to identify the specific spoilage organisms which were responsible for changes in pH and other organoleptic characteristics during deterioration of the quality of the meat.

7.4 Metagenomics

Over the last two decades, the decrease in the cost of whole genome sequencing has significantly increased accessibility to a large number of microbial genomes. The availability of this information has changed the nature of microbiology, as well as the way we study microbial ecology (Wooley et al. 2010). Scientists are now able to study and compare microbial genomes, side by side, which has opened up new fields of study including comparative genomics and systems biology. These areas of study are highly beneficial because microbes very rarely live in single species communities. Microbes most often interact with each other, as well as their environment, which could also potentially include a host organism (Wooley et al. 2010).

Metagenomics describes the study of genetic material recovered from environmental samples such as food, soil, and water. Metagenomics allows scientists to study the variety microbial organisms within a sample, even if the organisms are unculturable in the laboratory (Coughlan et al. 2015). This type of technology allows scientists to discover how microbes function in communities. For example, scientists now better understand the diversity and complexity of the microbial communities that reside in the human gastrointestinal tract, many of which were previously unable to be cultured in the laboratory by traditional methods (Ley et al. 2008; Wooley et al. 2010). This type of analyses can reveal the identity of a species present within a sample, as well as provide insight into the metabolic activities and functional roles of the microbes present within a sample population (Langille et al. 2013).

A typical metagenomics analysis begins with the isolation of high-quality microbial DNA from an environmental sample which should represent all species present within the sample, qualitatively and quantitatively (Coughlan et al. 2015). Metagenomic DNA is then directly sequenced and analyzed using bioinformatics to determine the functional traits of the microbial organisms which are present in the initial sample (Coughlan et al. 2015). Data are typically analyzed using computer software to infer phylogenetic relationships with existing genomic databases (Santiago-Rodriguez et al. 2016). Operational Taxonomic Units (OTUs) are similar sequences of DNA that serve as markers of species relatedness. The abundance of these OTUs are calculated to determine the taxonomy and diversity of the microbial community present within the sample (Kuczynski et al. 2012). These techniques can be applied to help identify specific organisms and enzymes involved in food spoilage, as well as novel enzymes from natural sources to aid in food processing reactions (Coughlan et al. 2015).

7.5 CRISPR-Cas

Clustered regularly interspaced short palindromic repeats (CRISPR) and CRISPR-associated sequences (Cas) make up the adaptive immune system of bacteria, which protects against invasive genomic elements (Barrangou and Marraffini 2015). CRISPR arrays confer immunological memory and surveillance mechanisms in bacteria, and *cas* genes encode effector proteins when the cell is under attack (Selle and Barrangou 2015). Immunity mediated by CRISPR-Cas is categorized into three molecular processes including acquisition, expression, and interference (Barrangou 2013; Barrangou and Marraffini 2015). Acquisition occurs when foreign genetic elements, such as bacteriophage DNA, enter a host cell. The foreign genetic element "samples" a novel spacer sequence from the host cell's chromosome using a copy and paste process, which leads to the formation of an additional CRISPR repeat-spacer unit within the host cell. Expression and RNA biogenesis occur in mature surveillance CRISPR RNAs (crRNAs), which comprise a portion of the "sample" CRISPR spacer sequence and define the target sequence in which the mobile genetic element is attempting to invade. Interference against mobile genetic elements is driven by Cas nucleases, which allow recognition of the target sequence. Cas nucleases perform cleavage of complementary DNA elements to interfere with the invasion of the mobile genetic element and are defined by the crRNA guide sequence (Selle and Barrangou 2015). Scientists have found a way to reprogram CRISPR-Cas system to edit and remodel the genomes of a variety of organisms.

It is believed that the development of the CRISPR-Cas system has reinvigorated the field of functional genomics in a scientific area that has recently been defined by whole genome sequencing (Selle and Barrangou 2015). Scientists are now using the CRISPR-Cas system to perform genome editing in a diverse range of organisms including eukaryotes and prokaryotes. Early application of CRISPR-Cas technology arose from food-science-driven research when industrial starter cultures for milk fermentation processes were characterized (Barrangou et al. 2007). The technology has also been applied to other organisms of interest across the field of food science including yeast, corn, rice, and tomatoes (Selle and Barrangou 2015). Other applications include genome editing for improvement of growth and disease resistance in food crops and animals (Selle and Barrangou 2015).

7.6 Foodomics

The term "foodomics" has been defined as a new discipline, which describes the study of food and nutrition through the application of advanced "omic" technologies to improve the confidence, health, and well-being of the consumer (Cifuentes

2009; Herrero et al. 2010; Herrero et al. 2012). Foodomics is regarded as a global discipline in which food, nutrition, advanced analytical techniques, and bioinformatic tools are brought together (Herrero et al. 2012; Andjelković et al. 2017). The suffixes "ome" and "omics" are derived from the term "genome," which was created by Hans Winkler in 1920 (Capozzi and Bordoni 2013). The suffix is now used to describe some of the major high-throughput "omics" methodologies including genomics, transcriptomics, proteomics, metabolomics, allergomics, lipidomics, nutrigenomics, as well as many others. These disciplines include the study of genes, RNA transcripts, proteins, metabolites, allergens, fats, and other nutrients that can be found within the cells of foods and food ingredients. Advanced technologies including mass spectrometry (MS)-based tools are commonly used to carry out "omics" based studies (Herrero et al. 2012). Other technologies including nuclear resonance spectroscopy and liquid chromatography (LC) are commonly used in metabolomics-based studies (Gibbons et al. 2015).

Foodomics is a systems science that can be used to aid in the development of transgenic foods, metabolic study for compound profiling, biomarker analysis related to food quality, the effects of food bioactivity on human health, as well as address food safety issues (Herrero et al. 2012; Quigley et al. 2012). These tools can provide a health assessment of plants and animals as food producers (García-Cañas et al. 2012), as well as the production and monitoring of food quality (Gašo-Sokač et al. 2011). Foodomics can also help scientists to further examine the rates and causes of change in foods during storage in order to extend shelf life. For example, food products including red wine (Arapitsas et al. 2016) and iceberg lettuce (Garcia et al. 2016) have been investigated using LC and MS platforms to determine the impact of storage specifications on their metabolic profile (Castro-Puyana et al. 2017; Bayram and Gökırmaklı 2018). Scientists have also used variety of foodomics techniques, including MC, LC, and genomics, to demonstrate that the shelf life of fruits can be enhanced by suppressing ripening-specific enzymes and slowing the rate of fruit softening (Meli et al. 2010).

7.7   Current Challenges and Research Needs

Novel approaches to enhance the production and shelf life of foods are warranted in order to feed the world's growing population. The emerging fields of science as described in this chapter can help to further our knowledge regarding the effects, relevance, and significance of the changes that occur in foods over time. The availability of advanced genomics and genetics tools will influence the integration of a mechanistic and evolutionary approach to tackling food quality and food safety obstacles (West et al. 2006; Sieuwerts et al. 2008). As the field of food science continues to progress, we are likely to see a movement from the study of the single components of food toward a systems approach, where we can better understand how all the components of a product form a complex network associated with specific biological functions (Santiago-Rodriguez et al. 2016). These approaches may be used more frequently, as the technology becomes more advanced and the cost continues to become more and more affordable.

References

Andjelković, U., M.Š. Gajdošik, D. Gašo-Sokač, T. Martinović, and D. Josić. 2017. Foodomics and food safety: Where we are. *Food Technology and Biotechnology* 55 (3): 290–307.

Arapitsas, P., A. Della Corte, H. Gika, L. Narduzzi, F. Mattivi, and G. Theodoridis. 2016. Studying the effect of storage conditions on the metabolite content of red wine using HILIC LC–MS based metabolomics. *Food Chemistry* 197: 1331–1340.

Barrangou, R. 2013. CRISPR-Cas systems and RNA-guided interference. *Wiley Interdisciplinary Reviews RNA* 4: 267–278.

Barrangou, R., and L.A. Marraffini. 2015. CRISPR-Cas systems: Prokaryotes upgrade to adaptive immunity. *Molecular Cell* 54: 234–244.

Barrangou, R., C. Fremaux, H. Deveau, M. Richards, P. Boyaval, S. Moineau, D.A. Romero, and P. Horvath. 2007. CRISPR provides acquired resistance against viruses in prokaryotes. *Science* 315: 1709–1712.

Bayram, M., and Ç. Gökırmaklı. 2018. Horizon scanning: How will metabolomics applications transform food science, bioengineering, and medical innovation in the current era of Foodomics? *OMICS* 22 (3): 177–183. https://doi.org/10.1089/omi.2017.0203.

Capozzi, F., and A. Bordoni. 2013. Foodomics: A new comprehensive approach to food and nutrition. *Genes & Nutrition* 8: 1–4.

Casas, V., and S. Maloy. 2014. Genomic and metagenomic approaches for predicting pathogen evolution. In *One Health: People, Animals, and the Environment*, ed. R.M. Atlas and S. Maloy, 227–235. Washington, DC: American Society for Microbiology.

Casas, V., and F. Rohwer. 2007. Phage metagenomics. *Methods in Enzymology* 421: 259–268.

Castro-Puyana, M., R. Pérez-Míguez, L. Montero, and M. Herrero. 2017. Application of mass spectrometry-based metabolomics approaches for food safety, quality and traceability. *Trends in Analytical Chemistry* 93: 102–118.

Cifuentes, A. 2009. Food analysis and foodomics. *Journal of Chromatography. A* 1216: 7109–7110.

Coughlan, L.M., P.D. Cotter, C. Hill, and A. Alvarez-Ordóñez. 2015. Biotechnological applications of functional metagenomics in the food and pharmaceutical industries. *Frontiers in Microbiology* 6: 672.

Davey, H.M. 2011. Life, death, and in-between: Meanings and methods in microbiology. *Applied and Environmental Microbiology* 77 (16): 5571–5576.

Ecrolini, D. 2013. High-throughput sequencing and metagenomics: Moving forward in the culture-independent analysis of food microbial ecology. *Applied and Environmental Microbiology* 79 (10): 3148–3155.

Fiers, W., R. Contreras, F. Duerinck, G. Haegman, D. Iserentant, et al. 1976. Complete nucleotide sequence of bacteriophage ms2 RNA: Primary and secondary structure of the replicase gene. *Nature* 260: 500–507.

Fleischmann, R.D., M.D. Adams, O. White, R.A. Clayton, E.F. Kirkness, et al. 1995. Whole-genome random sequencing and assembly of *haemophilus influenza* rd. *Science* 269: 496–512.

Floros, J., R. Newsome, and W. Fisher. 2010. Feeding the world today and tomorrow: The importance of food science and technology: An IFT Scientific Review. *Comprehensive Reviews in Food Science and Food Safety* 9: 572–599.

Garcia, C.J., R. García-Villalba, Y. Garrido, M.I. Gil, and F.A. Tomás-Barberán. 2016. Untargeted metabolomics approach using UPLC-ESI-QTOF-MS to explore the metabolome of fresh-cut iceberg lettuce. *Metabolomics* 12: 138.

García-Cañas, V., C. Simó, M. Herrero, E. Ibáñez, and A. Cifuentes. 2012. Present and future challenges in food analysis: Foodomics. *Analytical Chemistry* 48: 10150–10159.

Gašo-Sokač, D., S. Kovač, and D.J. Josić. 2011. Use of proteomic methodology in optimization and processing and quality control of food of animal origin. *Food Technology and Biotechnology* 49: 397–412.

Gibbons, H., A. O'Gorman, and L. Brennan. 2015. Metabolomics as a tool in nutritional research. *Current Opinion in Lipidology* 26: 30–34.

Gilbert, J.A., and C.L. Dupont. 2011. Microbial metagenomics: Beyond the genome. *Annual Review of Marine Science* 3: 347–371.

Gram, L., L. Ravn, M. Rasch, J.B. Bruhn, A.B. Christensen, and M. Givskov. 2002. Food spoilage-interactions between food spoilage bacteria. *International Journal of Food Microbiology* 78 (1–2): 79–97.

Handelsman, J., M.R. Rondon, S.F. Brady, J. Clardy, and R.M. Goodman. 1998. Molecular biological access to the chemistry of unknown soil microbes: A new frontier for natural products. *Chemistry & Biology* 5: R245–R249.

Heather, J.M., and B. Chain. 2016. The sequence of sequencers: The history of sequencing DNA. *Genomics* 107 (1): 1–8.

Herrero, M., V. García-Cañas, C. Simo, and A. Cifuentes. 2010. Recent advances in the application of CE methods for food analysis and foodomics. *Electrophoresis* 31: 205–228.

Herrero, M., C. Simó, V. García-Cañas, E. Ibáñez, and A. Cifuentes. 2012. Foodomics: MS-based strategies in modern food science and nutrition. *Mass Spectrometry Reviews* 31: 49–69.

Institute of Food Technologists (IFT). 2000. *Who Benefits from rDNA Biotechnology-Derived Foods?* http://www.ift.org/knowledge-center/read-ift-publications/science-reports/expert-reports/biotechnology-and-foods/who-benefits-from-rdna-biotechnology-derived-foods.aspx. Accessed 23 October 2017.

Kort, R., B.J. Keijser, M.P. Caspers, F.H. Schuren, and R. Montijn. 2008. Transcriptional activity around bacterial cell death reveals molecular biomarkers for cell viability. *BMC Genomics* 9: 590.

Kuczynski, J., J. Stombaugh, W.A. Walters, A. Gonzalez, J.G. Caporaso, and R. Knight. 2012. Using QIIME to analyze 16S rRNA gene sequences from microbial communities. *Current Protocols in Microbiology* Unitas 1E: 5.

Kumar, S., and A. Kumar. 2015. Role of genetic engineering in agriculture. *Plant Archives* 15: 1–6.

Langille, M.G.I., J. Zaneveld, J.G. Caporaso, D. McDonald, D. Knights, J.A. Reyes, et al. 2013. Predictive functional profiling of microbial communities using 16s rRNA marker gene sequences. *Nature Biotechnology* 31: 814–821.

Ley, R.E., M. Hamady, C. Lozupone, P.J. Turnbaugh, R.R. Ramey, J.S. Bircher, M.L. Schlegel, T.A. Tucker, M.D. Schrenzel, R. Knight, and J.I. Gordon. 2008. Evolution of mammals and their gut microbes. *Science* 320: 1647–1651.

Meli, V.S., S. Ghosh, T.N. Prabha, N. Chakraborty, S. Chakraborty, and A. Datta. 2010. Enhancement of fruit shelf life by suppressing N-glycan processing enzymes. *PNAS* 107 (6): 2413–2418.

Patz, J.A., P. Daszak, G.M. Tabor, A.A. Aguirre, M. Pearl, J. Epstein, N.D. Wolfe, A.M. Kilpatrick, J. Foufopoulos, D. Molyneux, D.J. Bradley, and Working Group on Land Use Change and Disease Emergence. 2004. Unhealthy landscapes: Policy recommendations on land use change and infectious disease emergence. *Environmental Health Perspectives* 112: 1092–1098.

Piotrowska-Cyplik, A., K. Myszka, J. Czarny, K. Ratajczak, R. Kowalski, R. Biegańska-Marecik, J. Staninska-Pięta, J. Nowak, and P. Cyplik. 2017. Characterization of specific spoilage organisms (SSOs) in vacuum-packaged ham by culture-plating techniques and MiSeq next-generation sequencing technologies. *Journal of the Science of Food and Agriculture* 97: 659–668.

Qin, C., C. Yu, Y. Shen, X. Fang, L. Chen, J. Min, et al. 2014. Whole-genome sequencing of cultivated and wild peppers provides insights into *Capsicum* domestication and specialization. *PNAS* 111 (14): 5135–5140.

Quigley, L., O. O'Sullivan, T.P. Beresford, R.P. Ross, G.F. Fitzgerald, and P.D. Cotter. 2012. High-throughput sequencing for detection of subpopulations of bacteria not previously associated with artisanal cheeses. *Applied and Environmental Microbiology* 78: 5717–5723.

Riesenfeld, C.S., P.D. Schloss, and J. Handelsman. 2004. Metagenomics: Genomic analysis of microbial communities. *Annual Review of Genetics* 38: 525–552.

Sanger, S.F., and A.R.C. Nicklen. 1977. DNA sequencing with chain-terminating. *Proceedings of the National Academy of Sciences* 74: 5463–5467.

Sanger, F., A.R. Coulson, T. Friedmann, G.M. Air, B.G. Barrell, et al. 1978. The nucleotide sequence of bacteriophage phix174. *Journal of Molecular Biology* 125: 225–246.

Santiago-Rodriguez, T.M., R. Cano, and R. Jiménez-Flores. 2016. Potential applications of metagenomics to assess the biological effects of food structure and function. *Food & Function* 7 (10): 4160–4169.

Selle, K., and R. Barrangou. 2015. CRISPR-based technologies and the future of food science. *Journal of Food Science* 80 (11): R2367–R2371.

Sieuwerts, S., F.A.M. de Bok, J. Hugenholtz, and J.E.T. van Hylckama Vlieg. 2008. Unraveling microbial interactions in food fermentations: From classical to genomics approaches. *Applied and Environmental Microbiology* 74 (16): 4997–5007.

Toledo-Arana, A., O. Dussurget, G. Nikitas, N. Sesto, H. Gvet-Revillet, D. Balestrino, E. Loh, J. Gripenland, T. Tiensuu, K. Vaitkevicius, M. Barthelemy, M. Vergassola, M.-A. Nahori, G. Soubigov, B. Regnault, J.-Y. Coppee, M. Lecvit, J. Johansson, and P. Cossart. 2009. The Listeria transcriptional landscape from saprophytism to virulence. *Nature* 459: 950–956.

Waite, J.G., J.M. Jones, and A.E. Yousef. 2009. Isolation and identification of spoilage microorganisms using food-based media combined with rDNA sequencing: Ranch dressing as a model food. *Food Microbiology* 26: 235–239.

Watson, J., and F. Crick. 1953. Molecular structure of nucleic acids. *Nature* 171: 709–756.

Watson, J.D., T.A. Baker, S.P. Bell, A. Gann, M. Levine, and R. Losick. 2008. *Molecular Biology of the Gene*. 6th ed. Cold Spring Harbor, NY: Cold Spring Harbor Laboratory Press.

West, S.A., A.S. Griffin, A. Gardner, and S.P. Diggle. 2006. Social evolution theory for microorganisms. *Nature Reviews. Microbiology* 4: 597–607.

Wooley, J.C., A. Godzik, and I. Friedberg. 2010. A primer on metagenomics. *PLoS Computational Biology* 6: 1–13.

Zallen, D.T. 2003. Despite Franklin's work, Wilkins earned his Nobel. *Nature* 425: 15.

Chapter 8
The Changing Shelf Life of Chilled, Vacuum-Packed Red Meat

John Sumner, Paul Vanderlinde, Mandeep Kaur, and Ian Jenson

8.1 Introduction

The Australian red meat industry has a long history of export to a large number of distant markets. The advent of vacuum packing made a significant change to the industry and its ability to supply high-quality product to these markets. This chapter describes the attempts to develop a solid Australian meat export trade prior to the development of vacuum packing, and then the incremental, but significant, improvements in shelf life since that time. While the attention of the industry, and this chapter, is on beef, comparisons are made to the shelf life of lamb.

In 2016, Australia exported red meat to more than 130 countries. The industry has come a long way from its origins, which began with seven Zebu cattle, 44 sheep, 19 goats and 32 pigs that landed with wobbly legs in 1788 after months at sea on the First Fleet which colonised Australia. This motley herd formed the basis of Australia's livestock industry, and quickly outgrew the consumption needs of the local population, leading to the need to export, almost entirely to England, the mother country.

The original version of this chapter was revised. The correction to this chapter is available at https://doi.org/10.1007/978-3-030-54375-4_9

J. Sumner
M&S Food Consultants, Deviot, TAS, Australia

P. Vanderlinde
Vanderlinde Consulting, Carbrook, QLD, Australia

M. Kaur
University of Tasmania, Hobart, TAS, Australia

I. Jenson (✉)
Meat & Livestock Australia, North Sydney, NSW, Australia
e-mail: ijenson@mla.com.au

© Springer Nature Switzerland AG 2021, Corrected Publication 2021
P. J. Taormina, M. D. Hardin (eds.), *Food Safety and Quality-Based Shelf Life of Perishable Foods*, Food Microbiology and Food Safety,
https://doi.org/10.1007/978-3-030-54375-4_8

A great deal is known about the early Australian meat industry, thanks to two definitive texts: *A Settlement Amply Supplied* by Dr. Keith Farrer (1980) and *A Review of Research since 1900* by Dr. Jim Vickery (1990). This introduction owes much to their work, which shows how problems with shelf life when shipping across both hemispheres were solved.

Mechanical refrigeration ('cold on demand') revolutionised the food industry, replacing an existing global trade in natural ice. For Australia, refrigeration offered commercial possibilities and, in 1873, the SS Norfolk was loaded with a trial shipment of frozen meat to England; the trial was unsuccessful when the circulating brine system failed.

In 1879, the SS Strathleven, fitted with mechanical refrigeration was loaded with mutton, beef and butter, which were then frozen on board. After a 64-day voyage the Strathleven arrived in London with a 34-tonne cargo in excellent condition. Organoleptic testing using an untrained consumer panel proceeded at a lunch on board the vessel for 150 tasters, and samples to Queen Victoria received the royal assent.

Problems caused by a lack of knowledge of how to thaw frozen meat meant the Australian trade soon took second place to chilled meat from Argentina and Paraguay. Successfully landed in London in 1907 after a 14-day voyage, their product was markedly superior to Australian frozen meat because there was no 'drip', and it attracted a price premium.

For the next 60 years, Australian research concentrated on trying to solve the problem of delivering chilled meat to distant markets. Various processes were investigated, and scientists from the forerunner of the Commonwealth Scientific and Industrial Research Organisation (CSIRO) worked on commercialising the effect of carbon dioxide on extending shelf life of chilled meat (Empey and Vickery 1933). The first trial shipment of chilled beef under carbon dioxide took place in 1933, with forequarters held in gas-tight cargo spaces. There were great problems due to gas leaks and the consignment was landed in the United Kingdom with incipient spoilage. Fitting out vessels with gas-tight storage rooms proved difficult, and World War 2 brought an end both to the export trade and its research and development.

However, the advent of flexible packaging in the 1960s allowed the chilled vacuum-packed (VP) meat trade to move towards commercial reality; vacuum packaging technology progressed quickly thereafter so that, by the 1980s, shelf life of around 14 weeks at -1 °C was considered the industry norm. More recently, storage trials carried out in Australia on VP beef primals (whole muscles) indicate shelf life of around 28 weeks (Small et al. 2012) and Canadian researchers have reported similar extended shelf life (Youssef et al. 2014).

In this chapter, we discuss possible reasons for this significant extension in shelf life of chilled VP beef primals. We begin the journey by focusing on early R&D in Australia and New Zealand, countries for which the technology presented obvious commercial advantages. A review by Nottingham (1982) presents the early state of the art led by researchers at CSIRO (e.g. Grau 1979) and the New Zealand Meat Research Institute (e.g. Gill and Newton 1979).

We then focus on the question of whether shelf life extension has resulted from improvements in hygiene and refrigeration, leading to a shift in microbial communities in the vacuum pack. We also seek to distinguish between a focus solely on bacterial numbers in evaluating end of shelf life and one of sensory testing as the primary consideration.

8.2 Sensory and Microbiological Quality of Chilled Red Meats

Shelf life of chilled meats is determined when product is considered unacceptable by any one of the following purchasers in the supply chain:

- *Brokers* or wholesalers, who move product to the next purchaser
- *Further processors*, who portion meat for retail or food service use
- *Retailers*, who display product for sale
- *Food service operators*, who prepare meat for consumption
- *Consumers*, who purchase meat for consumption

For the final consumer, there are three occasions when quality is assessed:

- *At purchase*, when the meat appearance is evaluated, particularly meat colour and amount of drip (exudate, also known as purge)
- *On opening*, when odour and whether the meat is slimy/sticky due to excessive bacterial growth are major criteria
- *On consumption*, when odour, flavour and texture are important criteria

More detailed criteria by which purchasers assess chilled product packed in various ways are presented in Table 8.1.

The relationship between the microbiological profile and sensory changes differs between meat packed under aerobic conditions and that packed under atmospheres containing high concentrations of CO_2 (20% or more) such as occurs in vacuum (VP) and modified atmosphere packs (MAP). Inside a vacuum package, residual oxygen is consumed due to ongoing muscle respiration and bacterial growth and CO_2 is produced. This results in an atmosphere of <1% O_2 at 20–40% CO_2 (Egan et al. 1988). Under aerobic storage, shelf life ends quickly (around 12 days at 0–2 °C) because of the growth of psychrotrophic, Gram-negative bacteria which become biochemically active, such as *Pseudomonas* and *Shewanella*. In VP and MAP meats stored close to 0 °C, shelf life is extended past 20 weeks because of the inhibition of Gram-negative spoilage bacteria by the high CO_2 concentration and the microbial communities selected for by the storage conditions: lactic acid bacteria (LAB), which are relatively biochemically innocuous.

Growth of LAB in VP meat was demonstrated by Egan (1983) with a typical sigmoid growth curve observed in VP primals stored at 0 °C (Fig. 8.1), where the population of LAB increased to a maximum of $\log_{10}$ 7–8 cfu/cm^2 after around 10 weeks. The LAB were a small proportion of the aerobic plate count (APC) initially, and came to dominate the APC within the first few weeks. Shelf life was deemed ended after around 14 weeks due to persistent off-odours when packs were opened.

Table 8.1 Sensory criteria of red meat through shelf life

Retail meat package	Positive quality attributes	Attributes at end of shelf life
Over-wrapped tray (non-aged meat)	Pink-red bloom Odour of fresh meat	Loss of bloom, brown discolouration Off-odours, off-flavours, slime
MAP—High oxygen, high CO_2	Pink-red bloom Odour of fresh meat	Loss of bloom, brown discolouration Off-odours, off-flavours, slime Excessive purge (drip)
Vacuum pack	Purple meat colour, tight pack Short-lived confinement odour Minimal purge (drip)	Unacceptable, persistent odour Meat discoloured (brown, grey or green) in intact pack Excessive purge (drip)

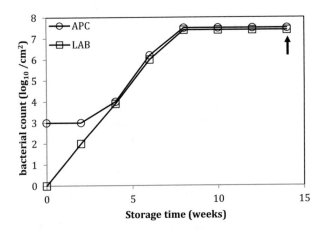

Fig. 8.1 Increase in the APC and LAB counts on VP beef primals stored at 0 °C (after Egan 1983), arrow shows time at which product was considered spoiled

These early studies indicate a complex relationship between the bacterial population in vacuum packs and the end of shelf life as evidenced by sensory evaluation; the attainment of maximum concentration per se was found not to equate with end of shelf life, one corollary being that it is the composition of the population and the build-up of metabolic end products which are critical in determining shelf life.

8.3 Then and Now: Sensory Aspects of Shelf Life

In the early 1970s, sensory panels recorded maximum acceptability of VP beef around 7 weeks, allowing the Japanese market to become a commercial reality. A total shelf life of 10–12 weeks at 0 °C was possible until ended by flavour changes described by panellists as sour, acidic and cheesy (Newton and Rigg 1979).

More recently, anecdotal evidence from meat exporters suggested that shelf life of VP beef has improved greatly (>20 weeks) compared with that found in the 1980s. To evaluate this suggestion, the CSIRO set up a trial on VP beef primals stored at −0.5 °C. At the expected end of shelf life (20 weeks), all samples were still of excellent sensory quality; no samples remained to determine the actual shelf life (Small et al. 2009).

A second, more extensive trial was set up using VP cube rolls and striploins sourced from six abattoirs located from a broad latitudinal range from Tasmania in the south (42°S) to far-north Queensland (19°S) to include stock from both temperate and subtropical areas. Vacuum packs were stored at −0.5 °C for up to 30 weeks. A trained sensory panel rated primals well for appearance and odour up to 28 weeks, after which there was a marked decline in score, particularly for striploins. Steaks cooked from striploins and cube rolls stored for 26 weeks scored well for aroma and flavour with sour flavours only noted at 28 weeks (Small et al. 2012).

In addition to documenting extraordinarily long shelf life, Small et al. (2012) recorded bacterial growth curves that did not conform with the traditional sigmoid curve comprising lag, exponential, stationary and decline phases, and the median counts barely reached 6 $\log_{10}$ cfu/cm^2. In addition, there were large variations in counts on each sampling day, often of the order of 4–5 $\log_{10}$ cfu/cm^2, both within and between processing establishments which had started with similar contamination levels.

The authors found it difficult to explain their findings, though they noted a New Zealand study of vacuum-packed beef striploins stored at −1 °C for 16 weeks in which the APC similarly did not follow the expected sigmoid curve, barely reaching 6 $\log_{10}$ cfu/cm^2 by the end of the trial (Penney et al. 1998). These observations are typical for growth of bacteria near their growth/no-growth boundary (McMeekin et al. 2002) where the lag phase prior to growth can vary considerably between samples.

In the trial of Small et al. (2012), the temperature in the storage rooms averaged −0.5 °C, cycling daily over the 30-week storage period between −2.7 °C and +1.8 °C. Near the minimum temperature, and at a meat pH of 5.4, a proportion of the community may be close to their growth: no-growth boundary. By contrast, near the maximum daily temperature and on meat of higher pH, a proportion of the community will be capable of growth.

Canadian researchers also report on extended shelf life of VP primals with striploins stored at either −1.5 °C or +2 °C being acceptable to taste panellists after 23 and 17 weeks, respectively (Youssef et al. 2014).

Thus, it can be concluded that shelf life of VP beef primals has increased markedly over recent years and, in the following sections, we follow the developments in process hygiene, packaging technology and product storage temperatures which have paralleled these increases.

8.4   Then and Now: Factors That Affect Shelf Life

Early reports of total bacterial levels were determined as the APC at 25 °C incubated for 96 h. These conditions allowed maximum recovery of slower growing organisms and organisms unable to grow at 30 °C or 35 °C, temperatures which are often used for bacterial counts. Historically, at packing, the APC of beef primal cuts centred around 3–4 $\log_{10}$ cfu/cm^2 (Egan 1983; Penney et al. 1998), with Egan et al. (1988) listing three prerequisites for optimal shelf life of VP meats:

- APC at packing no more than 2–3 $\log_{10}$ cfu/cm^2
- Packaging film with low oxygen permeability (≤ 25 cm^3/m^2/day at 23 °C)
- Good control of temperature throughout the storage period

These prerequisites offer a template against which current processing and storage of VP beef may be assessed.

8.4.1   Low Total Counts at Packing

To a large degree, the hygienic quality of beef cuts (primal and manufacturing meat) is linked with that of the carcasses from which they are derived. For Australian meat, it is possible to construct a 70-year comparative profile by utilising APC data from Grau (1979) and three national baseline studies (Table 8.2).

It should be emphasised that the data quoted by Grau (1979) were gathered at a single abattoir whereas baseline data reflect national average data. Nonetheless, the progressive reduction in APC appears to reflect the radical changes which the industry underwent beginning with the introduction of HACCP-based QA systems in the mid-1990s.

It is a reasonable assumption that improvements in the microbiological condition of the chilled carcass will be passed to the meat cuts fabricated from them and this is borne out by the most recent national baseline study of Australian primal cuts at the time of packing. As indicated in Table 8.3, the mean APC of two high-value primals, striploins and outsides, was <2 $\log_{10}$ cfu/cm^2 (Phillips et al. 2012), thus

Table 8.2 Beef carcass contamination in Australia, 1937–2004 (Sumner et al. 2011)

	$\log_{10}$ APC/cm^2	References
1937	3.88	Grau (1979)
1964	3.90	Grau (1979)
1978	2.79	Grau (1979)
1994	3.02	Vanderlinde et al. (1999)
1998	2.43	Phillips et al. (2001)
2004	1.33	Phillips et al. (2006)

Table 8.3 Aerobic plate counts (25 °C for 4 days) on Australian VP striploins and outsides (Phillips et al. 2012)

Primal cut	Concentration ($\log_{10}$ cfu/cm^2)		
	Mean	SD	Maximum
Striploin ($n = 572$)	1.3	1.0	5.3
Outside ($n = 572$)	1.5	1.0	4.2

meeting the first essential criterion for optimal shelf life attainment cited by Egan et al. (1988) of low total counts at the time of vacuum packing.

It seems clear, therefore, that improvements in process hygiene have occurred and have contributed to significantly lower levels of bacterial contamination on primals at the time of vacuum packing. Since signs of bacterial spoilage do not occur until some time after bacterial counts have reached their maximum, such improvements are likely to have had an impact on increasing shelf life of vacuum-packed meat over recent decades.

8.4.2   Low-Permeability Packaging Films

Technical information on vacuum packaging films and packaging technologies in the early days of VP chilled meats (Anon. 1970) reflects a technology in its infancy: impermeable films are described without any reference to oxygen transmission rates, and packaging technologies are broadly assigned to either *Evacuation and Sealing Without a Chamber*, or *Vacuum Sealed in a Chamber*.

An early edition of CSIRO's Meat Research News Letter (Anon 1971) describes the first steps in the export of chilled beef cuts citing the advantages of refrigerated containers and concluding: 'Vacuum packs in air-impermeable films extend the storage life by about two and a half times the period possible in air. They have the advantage of minimising weight loss while allowing ageing during this safe distribution life'.

The transmission rates for films routinely used for vacuum packaging have improved since the introduction of flexible film vacuum packing technology. Transmission rates in the range of 30–40 cm^3/m^2/day at 25 °C were reported in 1985 (Gill and Penney 1985) to 18.6 cm^3/m^2/day at 23 °C in recent years (Kiermeier et al. 2013). The multilayer structure of packaging materials has become more robust, plus tailored to fit specific cuts, thus reducing the likelihood of air entrapment or the development of leaks. Vacuum machines have also evolved, allowing rapid sealing of product (about 25 products/min) and form/fill machines are increasingly giving abattoirs the ability to manufacture retail-ready products in addition to the traditional primal/sub-primal in a vacuum bag.

Thus, meat in vacuum packs today is encased in superior packaging films with lower oxygen transmission, has reduced chance of leakage and is less likely to entrap oxygen, all contributing to greatly reduce the opportunity for the those microbes requiring oxygen or lower carbon dioxide concentration to grow and spoil product.

8.4.3 Temperature Throughout the Supply Chain

The optimum temperature for storage of meat was defined by Gill et al. (1988a) as -1.5 ± 0.5 °C, a temperature which minimises growth of spoilage organisms while preventing freezing of the product. The same workers also showed that small rises in temperature reduced shelf life significantly; at temperatures of 0 °C, 2 °C or 5 °C, the storage life was reduced by about 30%, 50% or 70%, respectively, compared with storage at -1.5 °C (Gill et al. 1988b).

Storage temperature is especially important in international trade where sea freight from Australasia to Europe may occupy up to 50 days, to which must be added time in the exporting country from packing to despatch, plus time required in the importing country for further processing, distribution and retailing.

Exporters routinely include data loggers in cartons of chilled VP primal cuts, a routine precaution in the event that shipments are delayed, or that the container temperature is higher than the -1 °C set point normally specified. A recent collation of time and temperature data for 135 sea voyages from Australia to markets in Europe, USA, Asia and the Middle East demonstrated that the mean temperature was usually close to 0 °C for the entire voyage, including brief temperature spikes during loading, unloading and trans-shipping. Occasionally, a higher temperature occurred which, on a long voyage, for example, to Europe, would prompt the importer to utilise the consignment as rapidly as possible (Sumner 2016).

Thus, it is likely that temperature control through the supply chain has improved over recent decades.

8.5 Then and Now: Microbial Communities on VP Beef Primals

Egan et al. (1988) considered three factors (product hygiene, film permeability and storage temperature) as important in achieving long shelf life but did not discuss microbial communities.

The selective effect of carbon dioxide against Gram-negative aerobes was established in the 1930s (Haines 1933) and applied in the VP meat context by Ingram (1962). As mentioned earlier, trade in chilled VP beef preceded a deep knowledge of the microbial communities which developed on the meat surface. However, work through the 1970s established that provided meat of normal pH (5.4–5.8) was packed in a film of low oxygen permeability (<100 cm^3/m^2/day at 25 °C) and an atmosphere of 20% carbon dioxide and $<1\%$ oxygen, lactic acid bacteria (LAB) would grow to be the dominant flora at the time of spoilage (Egan 1983). However, if the meat pH were higher than normal and/or the film allowed a less inhibitory $CO_2:O_2$ ratio, spoilage due to bacteria other than LAB could occur with *B. thermosphacta* causing early spoilage due to the presence of unacceptable dairy-like odours (Campbell et al. 1979). As well, if film permeability were relatively high (150 $cm^3/$

m²/day), *Shewanella putrefaciens* could spoil high pH meat due to production of hydrogen sulphide giving rise to rotten egg-type odours and greening of the meat. Finally, at higher than optimum storage temperatures, *Enterobacteriaceae* could grow on high-pH meat producing hydrogen sulphide and malodourous amines such as cadaverine and putrescine (Dainty et al. 1979).

Still in the early days of the new VP technology, the knowledge that spoilage was due to the predominance of lactic acid bacteria, particularly species of *Lactobacillus*, was based on the phenotypic and biochemical characteristics of organisms growing on selective media such as MRS agar. However, there were numerous reports (e.g. Hitchener et al. 1982) of 'unusual lactobacilli' from VP meat which were unable to grow on acetate agar and, on the basis of biochemical, physiological and chemical criteria, Collins et al. (1987) ascribed several atypical lactobacilli to a new genus, *Carnobacterium*.

More recent studies cite the predominant organisms in VP beef as *Carnobacterium divergens*, *C. piscicola*, *Lactobacillus sakei*, *L. curvatus*, *Leuconostoc gelidum*, *Leuc. carnosum* and *B. thermosphacta* (Ercolini et al. 2006; Fontana et al. 2006; Jones 2004; Nissen and Sorheim 1996; Sakala et al. 2002).

Modern genetic methods (metagenomics) provide opportunities to investigate the microbial communities and the possibility that the selection of certain bacteria may also affect shelf life. In a recent study, molecular techniques were used to further investigate the microbial communities extracted as rinsates from VP primals acquired by Small et al. (2012). Early in the shelf life, the microbial community on both cube rolls and striploins primal cut was dominated by Gram-negative, aerobic bacteria (*Pseudomonas* spp.). By week 16, various species of the lactic acid bacteria including *Carnobacterium* were detected and, by week 30, *C. maltaromaticum* or *C. divergens* had become dominant.

Carnobacteria have a number of positive effects on refrigerated meat (see review by Leisner et al. 2007) because they produce lactic acid, lowering the pH of meat, and some strains synthesise antibacterial compounds (bacteriocins), which may inhibit slower growing microorganisms. *C. divergens* and *C. maltaromaticum* have been studied extensively as protective cultures in order to inhibit growth of *Listeria monocytogenes* in fish and meat products and *C. maltaromaticum* has been shown to inhibit a wide range of bacterial isolates found in vacuum-packed beef (Zhang et al. 2015).

Recent work on VP beef primals has shown that storage temperature causes a shift in the profile of microbial communities in the vacuum pack. Low storage temperatures (≤ 2 °C) favoured the growth of lactic acid bacteria whereas storage at higher temperatures (4 °C or 8 °C) resulted in increased prevalence of non-LAB *Enterobacter*, *Serratia* and *Enterococcus*, bacterial species usually associated with meat spoilage (Ross et al. 2016).

The finding of significant differences in microbial communities during storage at 'low' versus 'high' temperatures is particularly relevant when exporting to markets where maintaining the integrity of the cold chain can be difficult (e.g. >40 °C is not unusual for much of the Middle Eastern year), or where storage and distribution infrastructure is not able to maintain the cold chain close to 0 °C.

8.6   Shelf Life of Vacuum-Packed Lamb

The shelf life of lamb is accepted as being shorter than beef, with the Australian industry often claiming about 12 weeks at −0.5 °C though this shelf life may be exceeded (Kiermeier et al. 2013). The shelf life of lamb has been reviewed by Mills et al. (2014), and among the reasons for its shorter shelf life are the following:

- Higher bacterial levels of lamb primals at packing (Phillips et al. 2013).
- Increased difficulty packing smaller primals, and bone-in primals.
- Lamb primals typically contain several muscle groups, most of which have a higher pH than beef primals.
- Most lamb primals contain both fat and lean surfaces, which leads to localised areas of high pH and less inhibition of microbes.

8.7   Conclusions

At the outset of this chapter, we undertook to pronounce on whether shelf life extension has resulted from improvements in hygienic processing, packing technology and the cold chain, leading to a shift in microbial communities in the vacuum pack.

Based on the evidence we have assembled, our conclusions are, firstly, that microbial communities have probably not changed over the decades, and that *Carnobacterium* has always dominated the microbial community on VP meat. Secondly, we conclude that extended shelf life of VP beef primals has been due to improvements in initial count, barrier film permeability, vacuum machines, and temperature control through storage and shipping.

On the other hand, there is much still to adduce about how conditions under vacuum affect communities. For instance, in the normal commercial setting of low-permeability packaging and storage below 0 °C, does pH become the key factor in determining the ultimate microbial community? If so, is this why Small et al. (2012) found some VP beef replicates with APCs of 7–8 $\log_{10}$ cfu/cm^2 while others on the same sampling day had $\log_{10}$ 3–4 cfu/cm^2? Is the higher pH of lamb primals, which contain a number of muscle groups, the reason why lamb shelf life is only about a half that of beef primals under the same storage conditions? Do variations of breeding lines of livestock impart variations of chemical structures of protein, collagen and/or fats that impact microbial metabolism and eventual organoleptic attributes of meat during shelf life? Is the proportion of psychrotrophic, CO_2-tolerant bacteria on cuts at the time of vacuum packing the critical microbial determinant of shelf life? These are all questions that need to be answered if we are to have a complete understanding of the factors that determine shelf life of VP meat.

References

Anonymous. 1970. *The use of packaging films for chilled fresh meats. Meat Research News Letter 70/6.* Brisbane, Australia: CSIRO Division of Food Research.

———. 1971. *Shipment of chilled beef cuts. Meat Research News Letter 71/2.* Brisbane, Australia: CSIRO Division of Food Research.

Campbell, R., A. Egan, F. Grau, and B. Shay. 1979. The growth of *Microbacterium thermosphactum* on beef. *Journal of Applied Bacteriology* 47: 505–509.

Collins, M., J. Farrow, B. Phillips, S. Ferusu, and D. Jones. 1987. Classification of *Lactobacillus divergens, Lactobacillus piscicola*, and some catalase-negative, asporogenous rod-shaped bacteria from poultry in a new genus, *Carnobacterium. International Journal of Systematic Bacteriology* 37: 310–316.

Dainty, R., B. Shaw, C. Harding, and S. Michanie. 1979. The spoilage of vacuum packed beef by cold tolerant bacteria. In *Cold tolerant microbes in spoilage and the environment*, SAB Technical Series, ed. A. Russell and R. Fuller, vol. 13. London: Academic.

Egan, A. 1983. Lactic acid bacteria of meat and meat products. *Antonie Van Leeuwenhoek* 49: 327–336.

Egan, A., I. Eustace, and B. Shay. 1988. Meat packaging—maintaining the quality and prolonging the storage life of chilled beef, pork and lamb. Proc 34th international congress of meat science and technology.

Empey, W., and J. Vickery. 1933. The use of carbon dioxide in the storage of chilled beef. *Journal of the Council of Scientific and Industrial Research* 6: 233–243.

Ercolini, D., F. Russo, E. Torrieri, P. Masi, and F. Villani. 2006. Changes in the spoilage-related microbiota of beef during refrigerated storage under different packaging conditions. *Applied and Environmental Microbiology* 72: 4663–4671.

Farrer, K. 1980. *A settlement amply supplied: Food technology in nineteenth century Australia.* Melbourne: Melbourne University Press.

Fontana, C., P. Cocconcelli, and G. Vignolo. 2006. Direct molecular approach to monitoring bacterial colonization on vacuum-packaged beef. *Applied and Environmental Microbiology* 72: 5618–5622.

Gill, C., and K. Newton. 1979. Spoilage of vacuum-packaged dark, firm, dry meat at chill temperatures. *Applied and Environmental Microbiology* 37: 362–364.

Gill, C., and N. Penney. 1985. Modification of in-pack conditions to extend the storage life of vacuum packaged lamb. *Meat Science* 14: 43–60.

Gill, C., D. Phillips, and J. Harrison. 1988a. Product temperature criteria for shipment of chilled meats to distant markets. In *Proc 1st International Refrigeration Conference*, 40–47. Brisbane: Refrigeration for Food and People.

Gill, C., D. Phillips, and M. Loeffen. 1988b. A computer program for assessing the remaining storage life of chilled red meats from product temperature history. In *Proc 1st International Refrigeration Conference*, 35–39. Brisbane: Refrigeration for Food and People.

Grau, F. 1979. Fresh meats: Bacterial association. *Archiv für Lebensmittelhygiene* 30: 81–116.

Haines, R. 1933. The influence of carbon dioxide on the rate of multiplication of certain bacteria as judged by viable counts. *Journal of Society of Chemical Industry* 52: 13T.

Hitchener, B., A. Egan, and P. Rogers. 1982. Characteristics of lactic acid bacteria isolated from vacuum-packaged beef. *Journal of Applied Bacteriology* 52: 31–37.

Ingram, M. 1962. Microbiological principles in prepacking meat. *Journal of Applied Bacteriology* 25: 259–281.

Jones, R. 2004. Observations on the succession dynamics of lactic acid bacteria populations in chill-stored vacuum-packaged beef. *International Journal of Food Microbiology* 90: 273–282.

Kiermeier, A., M. Tamplin, D. May, G. Holds, M. Williams, and A. Dann. 2013. Microbial growth, communities and sensory characteristics of vacuum and modified atmosphere packaged lamb shoulders. *Food Microbiology* 36: 305–315.

Leisner, J., B. Laursen, H. Prévost, D. Drider, and P. Dalgaard. 2007. *Carnobacterium*: Positive and negative effects in the environment and in foods. *FEMS Microbiological Reviews* 31: 592–613.

McMeekin, T., J. Olley, D. Ratkowsky, and T. Ross. 2002. Predictive microbiology: Towards the interface and beyond. *International Journal of Food Microbiology* 73: 395–407.

Mills, J., A. Donnison, and G. Brightwell. 2014. Factors affecting microbial spoilage and shelf life of chilled vacuum-packed lamb transported to distant markets: A review. *Meat Science* 98: 71–80.

Newton, K., and W. Rigg. 1979. The effect of film permeability on the storage life and microbiology of vacuum packed meat. *Journal of Applied Bacteriology* 47: 433–441.

Nissen, H., and O. Sorheim. 1996. Effects of vacuum, modified atmospheres and storage temperature on the microbial flora of packaged beef. *Food Microbiology* 13: 183–191.

Nottingham, P. 1982. Microbiology of carcass meats. In *Meat Microbiology*, ed. M. Brown, 13–65. London, New York: Applied Science Publishers.

Penney, N., R. Bell, and S. Moorhead. 1998. Performance during retail display of hot and cold boned beef striploins after chill storage in vacuum or carbon dioxide packaging. *Food Research International* 31: 521–527.

Phillips, D., J. Sumner, J. Alexander, and K. Dutton. 2001. Microbiological quality of Australian beef. *Journal of Food Protection* 64: 692–696.

Phillips, D., D. Jordan, S. Morris, I. Jenson, and J. Sumner. 2006. A national survey of the microbiological quality of beef carcasses and frozen boneless beef in Australia. *Journal of Food Protection* 69: 1113–1117.

Phillips, D., K. Bridger, I. Jenson, and J. Sumner. 2012. An Australian national survey of the microbiological quality of frozen boneless beef and beef primal cuts. *Journal of Food Protection* 75: 1862–1866.

Phillips, D., S. Tholath, I. Jenson, and J. Sumner. 2013. Microbiological quality of Australian sheep meat in 2011. *Food Control* 31: 291–294.

Ross, T., J. Bowman, M. Kaur, C. Kocharunchit, and M. Tamplin. 2016. *Microbial Ecology and Physiology. University of Tasmania Report (Meat & Livestock Australia project G.MFS.0289).*

Sakala, R., H. Hayashidani, Y. Kato, T. Hirata, Y. Makino, A. Fukushima, T. Yamada, C. Kaneuchi, and M. Ogawa. 2002. Change in the composition of the microflora, on vacuum-packaged beef during chiller storage. *International Journal of Food Microbiology* 74: 87–99.

Small, A., A. Sikes, and I. Jenson. 2009. Prolonged storage of chilled vacuum packed beef from Australian export abattoirs. *Proc: International Conference on Meat science and Technology, Copenhagen, August 2009*, pp.1294–1297. http://www.icomst2009.dk/fileadmin/documents/ICOMST_PS8_Final.pdf.

Small, A., I. Jenson, A. Kiermeier, et al. 2012. Vacuum-packed beef primals with extremely long shelf life have unusual microbiological profile. *Journal of Food Protection* 75: 1524–1527.

Sumner, J. 2016. *The impact of transport to Australia's distant markets on the shelf-life of beef and sheep primals. Australian meat processors corporation (AMPC) report 2106.1075, North Sydney, Australia.*

Sumner, J., A. Kiermeier, and I. Jenson. 2011. Verification of hygiene in Australian manufacturing beef processing–Focus on *Escherichia coli* O157. *Food Protection Trends* 31 (8): 514–520.

Vanderlinde, P., B. Shay, and J. Murray. 1999. Microbiological quality of Australian beef carcass meat and frozen bulk packed beef. *Journal of Food Protection* 61: 437–443.

Vickery, J. 1990. *Food science and Technology in Australia: A review of research since 1900.* Sydney: CSIRO Publishing.

Youssef, M., C. Gill, and X. Yang. 2014. Storage life at 2°C or −1.5°C of vacuum-packaged boneless and bone-in cuts from decontaminated beef carcass. *Journal of the Science of Food and Agriculture* 94: 3118–3124.

Zhang, P., J. Baranyi, and M. Tamplin. 2015. Interstrain interactions between bacteria isolated from vacuum-packaged refrigerated beef. *Applied and Environmental Microbiology* 81: 2753–2761.

Correction to: The Changing Shelf Life of Chilled, Vacuum-Packed Red Meat

John Sumner, Paul Vanderlinde, Mandeep Kaur, and Ian Jenson

Correction to:
Chapter 8 in: P. J. Taormina, M. D. Hardin (eds.),
Food Safety and Quality-Based Shelf Life of Perishable Foods,
Food Microbiology and Food Safety,
https://doi.org/10.1007/978-3-030-54375-4_8

The original version of this chapter was inadvertently published with incorrect name of the first author as John Summer, and the same has been corrected as John Sumner.

The updated online version of this chapter can be found at
https://doi.org/10.1007/978-3-030-54375-4_8

Index

© Springer Nature Switzerland AG 2021
P. J. Taormina, M. D. Hardin (eds.), *Food Safety and Quality-Based Shelf Life of Perishable Foods*, Food Microbiology and Food Safety,
https://doi.org/10.1007/978-3-030-54375-4

160

Printed in the United States
by Baker & Taylor Publisher Services